Springer Theses

Recognizing Outstanding Ph.D. Research

Aims and Scope

The series "Springer Theses" brings together a selection of the very best Ph.D. theses from around the world and across the physical sciences. Nominated and endorsed by two recognized specialists, each published volume has been selected for its scientific excellence and the high impact of its contents for the pertinent field of research. For greater accessibility to non-specialists, the published versions include an extended introduction, as well as a foreword by the student's supervisor explaining the special relevance of the work for the field. As a whole, the series will provide a valuable resource both for newcomers to the research fields described, and for other scientists seeking detailed background information on special questions. Finally, it provides an accredited documentation of the valuable contributions made by today's younger generation of scientists.

Theses are accepted into the series by invited nomination only and must fulfill all of the following criteria

- They must be written in good English.
- The topic should fall within the confines of Chemistry, Physics, Earth Sciences, Engineering and related interdisciplinary fields such as Materials, Nanoscience, Chemical Engineering, Complex Systems and Biophysics.
- The work reported in the thesis must represent a significant scientific advance.
- If the thesis includes previously published material, permission to reproduce this must be gained from the respective copyright holder.
- They must have been examined and passed during the 12 months prior to nomination.
- Each thesis should include a foreword by the supervisor outlining the significance of its content.
- The theses should have a clearly defined structure including an introduction accessible to scientists not expert in that particular field.

More information about this series at http://www.springer.com/series/8790

Hao Zhang

Imaging the Rupture Processes of Earthquakes Using the Relative Back-Projection Method

Theory and Applications

Doctoral Thesis accepted by
School of Earth and Space Sciences, Peking University,
Beijing, China

Springer

Author
Dr. Hao Zhang
Department of Geology and Geophysics
University of Utah
Salt Lake City, UT
USA

Supervisor
Zengxi Ge
School of Earth and Space Sciences
Peking University
Beijing
P.R China

ISSN 2190-5053 ISSN 2190-5061 (electronic)
Springer Theses
ISBN 978-3-662-55237-7 ISBN 978-3-662-55239-1 (eBook)
https://doi.org/10.1007/978-3-662-55239-1

Library of Congress Control Number: 2017958609

Printed on acid-free paper

This Springer imprint is published by Springer Nature
The registered company is Springer-Verlag GmbH Germany
The registered company address is: Heidelberger Platz 3, 14197 Berlin, Germany

Supervisor's Foreword

I am much honored to be invited to write a foreword for the dissertation for Dr. Hao Zhang. Hao is my first graduate student. I met him back in 2006 when I was a lecturer in the department of Geophysics, Peking University. I was frightened but excited when I receive a phone call from a colleague recommending Hao to continue his graduate study under my guidance. It was a challenge and an opportunity for me since it was the first time to advise a student.

Two years later, a destructive earthquake hit Sichuan, China. We selected the earthquake source dynamics as Hao's research topic. Although it is a new topic for both of us, we take it as a good opportunity and put our enthusiasm on it. After, we discussed the procedure for back-projection algorithm, Hao programmed it and obtained the rupture process of Wenchuan earthquake. After that, he applied it to several earthquakes including the 2010 M8.8 Chilean earthquake, the 2011 M9.0 Tohoku earthquake and the 2012 M8.6 Sumatra earthquake. The research results became the main part of this dissertation. The dissertation also included the basic principle and the implementation procedure of the back-projection method which is tested by both synthetic data and real data. The shortage of the method is also listed and partly solved. Thus, the book should be a good reference for people who want to apply the method to retrieve the rupture process of earthquakes.

Nevertheless, the dissertation also recorded the growth process of a student from zero to a good researcher. I appreciate the academic degree committee of Peking University to select this dissertation as one of the best Ph.D. dissertation in the year 2013. It is really a reward for his talent and the perseverance and untiring industry he spent on it.

Hao and I spent 5 years, working together on the methods and applications of the relative back-projection method. It was an important period for both of us to grow up. When I look back, the scene is just like yesterday. I should thank him for working with me and being patient, diligent, and creative.

For the dissertation, how to resolve high-frequency sources of earthquakes accurately and quickly, a very fundamental seismological question, is addressed, since the high-frequency sources account for the several damaging to building nearby the fault. The back-projection approach Hao used can resolve the rupture processes of the earthquakes even in two hours once the data for the regional arrays are available. Moreover, Hao has improved the method in a sense like relative relocation of earthquakes, which dramatically mitigates the artifacts commonly existing in the original back-projection imaging. This is very significant for real-time rupture imaging of the earthquakes and thus for reducing the seismic hazard.

Besides the earthquake mitigation, the back-projection method is a powerful, effective tool to investigate the earthquake physics. An era on great earthquakes has seemingly been entered since the 2004 Mw 9.2 Sumatra earthquake. The back-projection method has been widely applied to study the great earthquakes since then. New rupture features like frequency-dependence have been observed by Hao and others in the great earthquakes. Thus, as the great earthquakes keep occurring, the back-projection has a huge potential in understanding the earthquake source further.

Thanks to Springer, who provides the chance to publish the dissertation. I hope this is the new start for Dr. Zhang's academic career. I wish him a prosperous future.

P.R. China
October 2017

Zengxi Ge

Abstract

Quickly and effectively imaging rupture processes of large earthquakes not only plays a vital role in saving people's life and broadcasting tsunami warning, but also boosts the recognition of nature of faulting and the evolution of regional tectonics. The rupture processes of the earthquakes could be imaged by back-projecting seismic data recorded at generalized and regional seismic arrays. However, the traditional back-projection imaging suffers severely from "swimming" of a large high-frequency energy burst in the range direction of an array. In this book, a relative back-projection method, modified from the traditional method, is adopted to dramatically reduce this kind of "swimming" artifacts through identifying the rupture front by referring to the reference station at each time step, leading to resolve the rupture processes of earthquakes more accurately.

At first, the displacement representation theorem is introduced to demonstrate mathematically the feasible of beamforming, which is fundamental to the relative back-projection method. In the meanwhile, to explore the validating scope of the method as well as spatial and temporal resolutions, a series of synthetic tests are also carried out.

To verify the effectiveness of the relative back-projection method, May 12, 2008, Mw 7.8 Wenchuan, China earthquake is first investigated. The rupture process of the earthquake is imaged within the Longmen Shan Fault Zone. In order to remove artifacts, the multiplicity of amplitude ratio, coherence, and energy of beam is used as the discriminating factor. Moreover, after the focal mechanisms of point sources comprising of the earthquake are employed to correct the amplitude variations resulting from radiation pattern, two high-frequency sources radiating maximum energy are presented: one in a town named Yingxiu near the epicenter, and the other one nearby Beichuan.

Typically, a generalized seismic array has a limited azimuthal coverage. If only one array is available, resolving the radiation of high-frequency energy from the weak branch of the rupture is extremely difficult for a bilaterally rupturing earthquake. To address this issue, the source region of the 2010 Chilean earthquake is divided into two parts, cutting through the epicenter along the azimuth direction of the array. Next, the two regions are separately back-projected using the relative back-projection method. The back-projection results show that there exists a gap area of high-frequency sources ~70 km north of the epicenter, which could be aseismic or be ruptured smoothly, leading to lack of high-frequency seismic energy.

The megathrust earthquakes are pervasively observed to rupture in a frequency-dependent way. Using seismic data filtered in multiple frequency bands at the European network (EU), the relative back-projection method can be used to study the frequency-dependent rupture of the 2011 Mw 9.0 Tohoku, Japan earthquake. The imaging results show that the higher frequency band of data are radiated from sources located deeper on the subduction interface. This evidently reflects the depth-dependent characteristics of the asperity distribution on the causative fault. Besides the lateral extension of the rupture along the strike, the rupture propagates near the top of the fault during the first 30 seconds due to large rupture dimensions.

Besides the artifacts stemming from the large energy burst, depth phases could cause artifacts for a single array. To overcome the limitation of a single array, three seismic networks in different azimuths are adopted to image the extremely complex rupture of the 2012 Mw 8.6 Sumatra offshore earthquake. The imaging results showed that the earthquake ruptures multiple mutually–orthogonal or conjugate faults during ~120 s. The three imaged faults include two nearly parallel dextral faults with the east-southeast striking and one NNE trending sinistral fault in between. The rupture on the second causative fault first extends SSWwards and then NNEwards. The rupture velocities of the two segments differ significantly; this could be caused by the different rupture depths. In addition, interaction among the multiple causative faults is observed. After the failure of the first causative fault, increases of the static and dynamic Coulomb stresses trigger the rupture of the second causative fault, while the dynamic stress changes carried by the SH or Love waves radiated from the southern branch of the second fault trigger the rupture of the third causative fault.

Keywords Generalized array imaging · Beamforming · Rupture process · Relative back-projection method · Anti-earthquake emergency response · High-frequency earthquake source

Contents

Chapter 1
Introduction

Rupture of an earthquake is initiated by the nucleation at the hypocenter when the stress is stronger than the fault strength, extends outwards on the causative fault, accompanying with overcoming the friction force on the fault, and eventually ceases when the accumulated strain energy on the fault is totally consumed or a barrier appears to stop the rupture. In recent years, large earthquakes (Mw $\geq$ 7.8) frequently occurred in the world and caused enormous loss of people's lives and damaging of buildings. Thus, it is of great importance to quickly investigate the rupture processes of large earthquakes to mitigate these losses resulting from the earthquakes or tsunamis and landsides induced by the earthquakes.

Earthquakes typically generate seismic waves propagating in the three-dimensional heterogeneous earth as body waves (*P* and *S* waves) and close to the surface as surface waves (Rayleigh or Love waves), and scattering as coda waves. These waves are recorded as seismograms by the seismometers installed on the surface or the sea floor as ocean bottom seismometers (OBSs). The seismograms include both the information of the earthquake sources and the earth's interior. In the teleseismic view, the earth's structure could be simplified as a one-dimensional velocity model such as AK135 (Kennett et al. 1995). Given the one-dimensional velocity model, the information of the earth's structure can be described as Green's function (Kikuchi and Kanamori 1982), which is used to address the rupture processes of earthquakes by deconvolving the Green's function from the seismograms (Aki and Richards 2002).

The investigation of the earthquake sources highly depends on the high-quality seismic data. In a natural environment, the seismic signals are usually contaminated or even covered by the background noises such as those made by oceanic waters striking the shore or human activities. Especially, those noises have a significant effect on the seismograms recorded at a single seismic station in a sedimentary basin due to the amplification of the sediments to the noises. To extract the weak seismic signals greatly influenced by the noises, one can use a seismic array composed of multiple neighboring seismic stations to carry out beamforming or slant stack on the seismograms. Since these array techniques can significantly increase the signal-to-noise ratio

H. Zhang, *Imaging the Rupture Processes of Earthquakes Using the Relative Back-Projection Method*, Springer Theses,
https://doi.org/10.1007/978-3-662-55239-1_1

of the seismograms, they are broadly applied to detect underground nuclear explosions and to relocate teleseismic events (Rost and Thomas 2002).

The array techniques can also be exploited to study the rupture processes of large earthquakes. For the large earthquakes, the signal-to-noise ratio of the seismograms is large enough to discriminate the seismic signals from the background noises. Moreover, the large earthquakes typically have a very long duration up to 500 s for the 2004 Mw 9.2 Sumatra earthquake (Lay et al. 2005) and thus can be synthesized with a series of subevents (Ge and Chen 2008). However, seismic signals radiated by the latter-ruptured subevents are contaminated by those stemming from the earlier-ruptured subevents. Since the subevents spatially and temporally differ from each other, the signals from these subevents, recorded at one array, have distinct ray paths and parameters. Given a frequency of the waveforms, which are decomposed in the wavenumber domain, the seismic array is typically sensitive to a wave field with a certain wavenumber; this indicates the seismic array can be treated as a frequency-wavenumber filter.

Generalized array techniques are modified from the aforementioned array techniques, but are applied to regional seismic networks instead of local, dense arrays with the same seismometers. Although stations composed of a regional seismic network have different instrumental seismometers, the seismograms are similar to those recorded at a traditional seismic array after the instrumental responses of the different seismometers are removed from the waveforms. Moreover, after the Global Position System (GPS) is pervasively used, the accuracy of timing for the stations in the regional seismic network is high enough for the traditional array techniques. More importantly, the number of the deployed seismic station nearly linearly increases with time all over the continents and gradually increases on the sea floor. For example, the USarray in North America and Hi-net in Japan have a station spacing less than 100 km. These two seismic networks are the typical generalized seismic arrays since the waveforms recorded at stations of the two seismic networks are similar enough to suppress the coda of the earlier-ruptured subevents to resolve the latter-ruptured subevents. Thus, all the seismic networks with small station spacing and large number of stations can be used as generalized seismic arrays to investigate the rupture processes of the earthquakes.

The array techniques are fundamental to array seismology. These techniques are first introduced. Next, the generalized array techniques used in the research on the earthquake sources are introduced in details. The issues encountered in the application of these techniques are also elaborated. For these issues, the ways to address them are demonstrated. Finally, the outline of the book is presented at the end.

1.1 Seismic Array and Array Techniques

An array is a receiver system comprising enough number of instruments distributing in a regular/certain geometry. This kind of instrumental arrays has been broadly applied in astronomy, communication and seismology. In the early twentieth century, anneal arrays were already applied in communication and radio-electric astronomy. The Partial Test Ban Treaty (PTBT), signed in 1963, only allowed the nuclear explosions conducted underground. Soon after the treaty, several seismic arrays were deployed and used to detect the global underground nuclear explosions. The arrays were composed of ~100 seismometers with an aperture of ~100 km such as Large Aperture Seismic Array (LASA) in Montana, USA and Norwegian Seismic Array (NORSAR) (Ringdal and Husebye 1982). Thus, the seismic array is defined as a receiver system composed of at least three stations with the same seismometers and having a spacing less than 1 km. Moreover, the waveforms recorded at the stations are transported to computers to conduct online detection and relocation on the regional or USarray events in a real-time way. Additionally, the data are restored to facilitate the data communication and further investigations on the detected events (Ringdal and Husebye 1982; Schweitzer et al. 2002).

Seismic arrays can be used to remotely detect low-yield underground nuclear explosions. The seismic signals generated by these nuclear explosions are dramatically weak and could even be covered by the background noises for a single station. However, these weak signals could be extracted by beamforming the waveforms recorded at the seismic arrays. For example, there occurred a nuclear explosion in east Kazakhstan, which had a very low yield of 0.1 ton TNT. The short-period signals radiated by the explosion were recorded by a seismic array (Alice Springs Array, ASAR) in Australia, which was composed of 20 stations and ~88° far away from the explosion. The short-period signals have been amplified to be above the background noises by beamforming the waveforms at the ASAR and can be clearly seen in the stacked waveforms (Douglas et al. 1999). Through further analysis on the ratio of body and surface waves of the beamforming/stacked signals, the source of the signals can be discriminated to be an underground nuclear explosion or not (Marshall and Basham 1972). Moreover, using the array techniques such as frequency-wavenumber analysis (Capon 1969, 1973), the back azimuth of wave impingement and apparent velocity can be estimated. Given a velocity model like AK135 (Kennett et al. 1995), the source of the signals can be relocated with an uncertainty less than 50 km (Ringdal and Husebye 1982).

As briefly mentioned above, the seismic arrays are broadly applied in detection of nuclear explosions, relocation of teleseismic earthquakes, and high-resolution structure of deep earth due to the high effectiveness of the array techniques to extract weak signals. Since the traditional seismic arrays have apertures less than 200 km (Frosch and Green 1966) and the epidistances of the arrays are larger than 30°, the impinging waves are assumed to be plane waves in most of the array techniques. Provided that the ray parameter or horizontal slowness of the impinging plane waves is known, arrival time delays of the station elements relative to a reference station can be thereby estimated. The principle of the array techniques is to back-project the waveforms with the time delays and then to sum up the back-projected waveforms to the stacked waveform. This means that only if the given ray parameter corresponds to that from the earthquake source, the power of the stacked waveforms after the back-projection will reach the maximum. Thus, the maximum power can be used as the criteria to determine the back azimuth and apparent velocity of the earthquake source. All the array techniques are conducted under the principle in a beamforming or reverse-time stacking way (Douglas 1998; Rost and Thomas 2002; Schweitzer et al. 2002).

There are two major kinds of array techniques: one in time domain, and the other one in frequency domain. The array techniques in time domain are called beam-packing (Schweitzer et al. 2002), and those in frequency domain are called frequency-wavenumber analyses (Capon 1969, 1973; Kværna and Doornbos 1986; Gupta et al. 1990). The difference between the two kinds of methods is the way to realize the reverse of the time delays, which is achieved by phase shift in frequency domain. However, there are no significant differences between results derived from the two kinds of methods using the same dataset. Additionally, relative to the linear stacking, Nth-root stacking suppresses the background noises better. Moreover, the signal-to-noise ratios (SNR) of the waveforms recorded at each array element are different, so the SNR of each element can be used as the weight to emphasize the stations having higher quality waveforms.

The response of a seismic array to seismic signals depends on the array transfer function (ATF). The ATF is determined by the array aperture, station spacing, number of stations, and array geometry (Haubrich 1968; Harjes 1990; Rost and Thomas 2002). This indicates the ATF is a deterministic property of the array, which is clear after the array is deployed. With the ATF, one can judge whether the imaging results have spatial alias or not; this facilitates the interpretation on the resultant imaging results.

1.2 Investigation on the Rupture of Earthquakes and Rupture Imaging Using Generalized Arrays

Rice (1980) divided the rupture mechanism of the earthquakes into three fields. The first field is in term of elastic wave field generated by an earthquake, which is the content of the classical seismology. For instance, the wave field from a double couple point source can be synthesized by convolving the Green's function of the elastic media with a source time function in a finite duration. With the seismograms recorded at the stations, the slip model on the causative fault can be inverted by fitting the observations with the synthesized wave field. This is the principle of the broadly applied finite fault model used to investigate the earthquake sources. The second field is to describe the rupture in the rocks. This field hypothesizes that earthquakes are caused by the abrupt fracture of the rocks when the loading stress is larger than the rock strength; this indicates earthquakes can be depicted by the failure mechanics. The third field focuses on the loading process of stress and argues that the earthquakes result from the large-scale tectonic deformation. The three fields view the earthquakes in three different perspectives; this leads to the different methods used in the investigation. However, the three fields are tied closely to each other. The progress in one field would promote the investigations in the other two fields. In this book, the rupture processes of the earthquake in the first field are the research objectives.

The first study on kinematic earthquake source is the elastic rebound theory proposed by Reid (1910, 1911) after he carried out a geological survey on the surface failure traces of the 1906 Las Angles earthquake. The kinematic earthquake source studies started in the middle of 1970s (Langston 1967; Burdick and Mellman 1976) and gradually became mature in early 1980s (Boatwright 1980; Kikuchi and Kanamori 1982; Olson and Apsel 1982; Hartzell and Heaton 1983). The significant increase of the computational ability, high-quality data collection, and inversion methods lead to the kinematic slip model being a powerful, effective tool to study the earthquake rupture.

However, there exist several issues in the inversion of the waveforms for the kinematic slip model. First, the solution of the inversion is not typically unique, because after the parameterization, the unknown variables are commonly larger than the observations, even if some constraints have been added in the inversion. Wagner and Langston (1989) applied body waves fitting to quantify the earthquake source and found that even if the slip distribution, source time function, and focal depth were changed, the synthesized waveforms from different slip models could fit the observations equally well. This could be caused by the trade-off effect between two variables such as the rupture velocity and slip rate on the fault patch (Spudich and Frazer 1984). Second, the constraints are added to address the issue of non-uniqueness of solution to the inversion. This could bring some subjective into the final slip model (Lay et al. 2010). Third, the error of the slip model could increase due to the inaccurate velocity model, the regularization to converge the solution, and predefined inaccurate fault dimension.

As global seismic stations increase, the generalized array techniques originally applied in local, dense arrays are gradually used for the regional seismic networks without the plane wave assumption. Compared with the aforementioned finite fault modeling, the imaging of the earthquake source using the generalized array techniques has advantages as follows. First, generalized technique is an imaging technique but is not involved in inversion; this leads to the unique imaging result. Second, the geometry of the fault does not need to be prescribed for the investigation using generalized array techniques. Third, no Green's function is required and this greatly reduces the computational time to obtain the rupture process of an earthquake. The high-frequency sources during the earthquake can be quickly identified. Since the regions nearby the high-frequency sources suffer more damaging, the quick determination of the high-frequency sources will provide very useful and important information for earthquake rescue response or even tsunami or landslide warning. Fourth, the waveforms used by the generalized array techniques have higher frequency than the finite fault modeling (Zhang et al. 2017). This broadens the frequency of data used in the earthquake source and provides a comprehensive recognition on the earthquake. Fifth, the rupture parameters such as fault dimension and rupture velocity can be used by the finite fault model; this constrains the slip model derived by the finite fault model better (Lay et al. 2010).

The first study on the rupture of earthquakes using the array techniques was conducted by Bolt et al. (1982). They employed the frequency-wavenumber method to investigate the 1981 Mb 6.9 earthquake 30 km northwest of a strong motion array SMART1 with an aperture of 2 km in Taiwan, China. The results showed that back azimuth of the impinging P waves became narrower with time, indicating that the rupture of the earthquake extended from north to south, though the detailed rupture was unclear. Similarly, the same frequency-wavenumber method was exploited to study the 1979 Imperial Valley earthquake by Spudich and Cranswick (1984) using waveforms recorded at a 213-meter-long linear array. The slowness vector of the impinging wave field radiated from the earthquake 5.6 km away from the center of the array varied with time. According to the evolution of the slowness vector with time, they obtained the rupture fronts changing with time on the fault. The array techniques were used to imaging the rupture of the earthquakes in a traditional way until 2004 (Goldstein and Archuleta 1991; Fletcher et al. 1992; Huang 2001; Almendros et al. 2002). The 2004 Mw 9.2 Sumatra–Andaman earthquake was the first earthquake studied by back-projection P waves recorded at generalized seismic arrays by Ishii et al. (2005). Since then, the back-projection method has been broad-widely applied to image the rupture processes of earthquakes and a lot of new knowledge on the earthquake source were gained from these back-projection studies (Walker et al. 2005; Allmann and Shearer 2007; Ishii et al. 2007; Kao et al. 2007; Liu et al. 2007; Honda et al. 2009; Du et al. 2009; Xu et al. 2009; Walker and Shearer 2009; Zhang and Ge 2010, 2012, 2014, 2016, 2017; Liu et al. 2010; Yao et al. 2011, 2012, 2013; Meng et al. 2011; Kiser and Ishii 2011; Koper et al. 2011a, b; Zhang et al. 2011, 2012, 2017; Wang and Mori 2011; Wang et al. 2012; Kiser and Ishii 2012; Sufri et al. 2012; Roten et al. 2012; Yagi et al. 2012; Maercklin et al. 2012; Meng et al. 2012a, b, c; Satriano et al. 2012; Yue et al. 2012). In the meanwhile, the traditional array techniques were still used in studying earthquake sources (Kruger and Ohrnberger 2005; Fletcher et al. 2006). However, the back-projection imaging results severely suffered from artifacts caused by depth phases (pP and sP) and initial time correction (Zhang et al. 2017). If not treated correctly, the artifacts would be explained to be rupture subevents; this leads to wrong interpretation on the rupture of earthquakes. Therefore, the back-projection method by Ishii et al. (2005) may need improvement to get rid of or at least mitigate the artifacts to make the imaging results more reliable.

1.3 Issues in the Back-Projection Studies

Although back-projection imaging has been broadly used in earthquake source, there still exist several issues in the back-projection imaging. The major issue of the back-projection imaging comes from artifacts. The artifacts have following sources. First, the depth phases from the same subevent could affect the direct P waves radiated from the latter subevents. Second, signals from larger subevents would influence those from smaller subevents due to trade-off between the rupture time and spatial locations along the ray path; this leads to "swimming" effect in imaging, which will be discussed below in details. Third, the initial time correction to the first subevent applied to other subevents would cause time bias due to the three-dimensional heterogeneous structure of the earth's interior.

Among these artifacts, "swimming" effect in imaging is very prominent (Xu et al. 2009; Meng et al. 2012a). The phenomenon is specifically observed as an imaged subevent changes the location with time toward/away from the array. One approach to mitigate this kind of artifacts is to use a reference time window (Meng et al. 2012a). This approach is similar to the relative back-projection method proposed by Zhang and Ge (2010), which can effectively mitigate the "swimming" artifacts. Another approach is to increase the back azimuth coverage of data with a global seismic array (Xu et al. 2009). This is similar to the case of the earthquake relocation, in which the full coverage in back azimuth could make the relocation of earthquakes accurate. However, not only the arrival times but also the waveforms were used in the back-projection. The waveforms change with back azimuth due to radiation pattern resulted from focal mechanism of earthquakes, rupture directivity, and heterogeneous structure of the earth's interior. This could lead to other kind of artifacts due to the non-similar waveforms caused by the factors described above (Koper 2016, personal communication).

1.4 Motivation and Objectives of the Study

As shown above, the tradition back-projection imaging suffers significantly from the "swimming" artifacts as well as other kinds of artifacts. In this study, we focus on these artifacts and mitigate them by improving the traditional back-projection method (Ishii et al. 2005). Thus, we have three objectives as follows. First, we proof the effectiveness of the generalized array techniques in a mathematical way. Second, we propose a new relative back-projection method to mitigate the artifacts, especially the "swimming" artifacts, in traditional back-projection imaging. Third, we apply the relative back-projection to image the rupture processes of several large earthquakes to demonstrate the effectiveness of the method and to broaden one's knowledge on the rupture of earthquakes.

1.5 Outline of the Book

To achieve the objectives described above, we outline the book as follows.

- Chapter 1 introduces background and current status of the investigation on the rupture processes of earthquakes, especially using the back-projection method, and briefly introduces the motivation and objectives of the relative back-projection imaging.

- Chapter 2 proofs the effectiveness and feasibility of the generalized array techniques in theory, introduces several widely used generalized array approaches, tests the relative back-projection approach with synthesized waveforms, and does a comparison of these results with those derived by the traditional back-projection method.
- Chapter 3 shows an application of the relative back-projection method (RBPM) to image the rupture process of the May 12, 2008, Mw 7.8 Wenchuan earthquake. In this chapter, the principle and procedure of the method are first introduced. Next, several operations are conducted to remove the artifacts in the imaging results. Finally, the imaging results are analyzed, and the relationship between high-frequency sources and damaging zones in the source region is discussed.
- Chapter 4 elaborates another application of the RBPM to the February 27, 2010, Mw 8.8 Chilean megathrust earthquake, which ruptures bilaterally. For the bilaterally rupturing earthquake, we apply a special strategy to image the earthquake. The rupture properties and seismic gap are discussed at the end.
- Chapter 5 elaborates another application of the RBPM to the March 11, 2011, Mw 9.0 Tohoku, Japan megathrust earthquake. For the unexpected earthquake, extraordinary rupture properties such as segmentation, frequency-dependent, and rupture velocity related to the structure of the subduction interface are observed by the RBPM.
- Chapter 6 demonstrates another application of the RBPM to April 11, 2012, Mw 8.6 Sumatra offshore earthquake. The earthquake occurred in Wharton basin, which is characterized by orthogonal or conjugate faults. The back-projection imaging shows that the earthquake does rupture multiple orthogonal or conjugate faults. Moreover, the latter-ruptured fault may be triggered by the rupture of the earlier-ruptured faults.
- Chapter 7 includes a discussion, summary and perspective on the RBPM, and rupture of the earthquakes.

References

Aki, K., and P.G. Richards, 2002. Quantitative seismology. University Science Books.

Allmann, B.P., and P.M. Shearer. 2007. A high-frequency secondary event during the 2004 Parkfield earthquake. *Science* 318 (5854): 1279–1283.

Almendros, J., B. Chouet, and P. Dawson. 2002. Array detection of a moving source. *Seismological Research Letters* 73 (2): 153–165.

Boatwright, J. 1980. A spectral theory for circular seismic sources; simple estimates of source dimension, dynamic stress drop, and radiated seismic energy. *Bulletin of the Seismological Society of America* 70 (1): 1–27.

Bolt, B.A., Y.B. Tsai, K. Yeh, and M.K. Hsu. 1982. Earthquake strong motions recorded by a large near-source array of digital seismographs. *Earthquake Engineering and Structural Dynamics* 10 (4): 561–573.

Burdick, L.J., and G.R. Mellman. 1976. Inversion of the body waves from the Borrego Mountain earthquake to the source mechanism. *Bulletin of the Seismological Society of America* 66 (5): 1485–1499.

Capon, J. 1969. High-resolution frequency-wavenumber spectrum analysis. *Proceedings of the IEEE* 57 (8): 1408–1418.

Capon, J. 1973. Signal processing and frequency-wavenumber spectrum analysis for a large aperture seismic array. *Method in Computational Physics.* 13: 1–59.

Douglas, A. 1998. Making the most of the recordings from short-period seismometer arrays. *Bulletin of the Seismological Society of America* 88 (5): 1155–1170.

Douglas, A., D. Bowers, P.D. Marshall, et al. 1999. Putting nuclear-test monitoring to the test. *Nature* 398 (6727): 474–475.

Du, H.L., L.S. Xu, and Y.T. Chen. 2009. Rupture process of the 2008 great Wenchuan earthquake from the analysis of the Alaska array data. *Chinese Journal of Geophysics* 52 (2): 372–378. (in Chinese).

Fletcher, J.B., L.M. Baker, P. Spudich, P. Goldstein, J.D. Sims, and M. Hellweg. 1992. The USGS Parkfield, California, dense seismograph array: UPSAR. *Bulletin of the Seismological Society of America* 82 (2): 1041–1070.

Fletcher, J.B., P. Spudich, and L.M. Baker. 2006. Rupture propagation of the 2004 Parkfield, California, earthquake from observations at the UPSAR. *Bulletin of the Seismological Society of America* 96 (4B): 129–142.

Frosch, R.A., and P.E. Green. 1966. The concept of a large aperture seismic array. *Proceedings of the Royal Society of London, Series A: Mathematical and Physical Sciences* 290 (1422): 368–384.

Ge, Z.X., and X.F. Chen. 2008. Point source stacking method to compute synthetic seismogram of finite moving planar source. *Acta Scientiarum Naturalium Universitatis Pekinensis, Beijing University (Natural Science)* 44: 407–412.

Goldstein, P., and R.J. Archuleta. 1991. Deterministic frequency-wavenumber methods and direct measurements of rupture propagation during earthquakes using a dense array: Theory and methods. *Journal of Geophysical Research* 96 (B4): 6173–6185.

Gupta, I.N., C.S. Lynnes, and R.A. Wagner. 1990. Broadband F-k analysis of array data to identify sources of local scattering. *Geophysical Research Letters* 17 (2): 183–186.

Harjes, H.P. 1990. Design and siting of a new regional array in Central Europe. *Bulletin of the Seismological Society of America* 80 (6B): 1801–1817.

Hartzell, S.H., and T.H. Heaton. 1983. Inversion of strong ground motion and teleseismic waveform data for the fault rupture history of the 1979 Imperial Valley, California, earthquake. *Bulletin of the Seismological Society of America* 73 (6A): 1553–1583.

Haubrich, R.A. 1968. Array design. *Bulletin of the Seismological Society of America* 58 (3): 977–991.

Honda, R., and S. Aoi. 2009. Array back-projection imaging of the 2007 Niigataken Chuetsu-oki earthquake striking the world's largest nuclear power plant. *Bulletin of the Seismological Society of America* 99 (1): 141–147.

Huang, B.S. 2001. Evidence for azimuthal and temporal variations of the rupture propagation of the 1999 Chi-Chi, Taiwan Earthquake from dense seismic array observations. *Geophysical Research Letters* 28 (17): 3377–3380.

Ishii, M., P.M. Shearer, H. Houston, and J.E. Vidale. 2005. Extent, duration and speed of the 2004 Sumatra-Andaman earthquake imaged by the Hi-Net array. *Nature* 435 (7044): 933–936.

Ishii, M., P.M. Shearer, H. Houston, and J.E. Vidale. 2007. Teleseismic P wave imaging of the 26 December 2004 Sumatra-Andaman and 28 March 2005 Sumatra earthquake ruptures using the Hi-net array. *Journal of Geophysical Research: Solid Earth (1978–2012)* 112: B11307.

Kao, H., S.J. Shan, G. Rogers, and H. Dragert. 2007. Migration characteristics of seismic tremors in the northern Cascadia margin. *Geophysical Research Letters* 34(3), L03304, doi: 10.1029/2006GL028430.

Kennett, B.L.N., E.R. Engdahl, and R. Buland. 1995. Constraints on seismic velocities in the earth from traveltimes. *Geophysical Journal International* 122 (1): 108–124.

Kikuchi, M., and H. Kanamori. 1982. Inversion of complex body waves. *Bulletin of the Seismological Society of America* 72 (2): 491–506.

Kiser, E., and M. Ishii. 2011. The 2010 Mw 8.8 Chile earthquake: Triggering on multiple segments and frequency-dependent rupture behavior. *Geophysical Research Letters* 38 (7), L07301, doi: 10.1029/2011GL047140.

Kiser, E., and M. Ishii. 2012. The March 11, 2011 Tohoku-oki earthquake and cascading failure of the plate interface. *Geophysical Research Letters* 39, L00G25.

Koper, K.D., A.R. Hutko, and T. Lay. 2011a. Along-dip variation of teleseismic short-period radiation from the 11 March 2011 Tohoku earthquake (Mw 9.0). *Geophysical Research Letters* 38: L21309.

Koper, K.D., A.R. Hutko, T. Lay, et al. 2011b. Frequency-dependent rupture process of the 2011 M (w) 9. 0 Tohoku Earthquake: Comparison of short-period P wave backprojection images and broadband seismic rupture models. *Earth, Planets and Space* 63 (7): 599–602.

Kruger, F., and M. Ohrnberger. 2005. Tracking the rupture of the Mw & equals; 9.3 Sumatra earthquake over 1,150 km at teleseismic distance. *Nature* 435 (7044): 937–939.

Kværna, T., and D. Doornbos. 1986. An integrated approach to slowness analysis with arrays and tree component stations. NORSAR semiannual technical summary, 1 October 1985–31 March 1986, NORSAR, Kjeller, Norway. Tech. Rep., Scientific Report.

Langston, C.A. 1967. A body wave inversion of the Koyna, India, earthquake of December 10, 1967, and some implications for body wave focal mechanisms. *Journal of Geophysical Research* 81 (14): 2517–2529.

Lay, T., H. Kanamori, C.J. Ammon, et al. 2005. The great Sumatra-Andaman earthquake of 26 December 2004. *Science* 308 (5725): 1127–1133.

Lay, T., C.J. Ammon, A.R. Hutko, and H. Kanamori. 2010. Effects of kinematic constraints on teleseismic finite-source rupture inversions: Great Peruvian earthquakes of 23 June 2001 and 15 August 2007. *Bulletin of the Seismological Society of America* 100 (3): 969–994.

Liu, N., Q.F. Chen, F.L. Niu, and Y. Chen. 2007. Imaging the rupture process of the 2004 Sumatra-Andaman earthquake using direct P waves. *Science Bulletin* 52(11): 1312–1316.

Liu, N., F.L., Niu, Q.F. Chen, and Y. Chen, 2010. Imaging the rupture of the 2010 M8.8 Chile earthquake with a broadband seismic array. *Chinese Journal Geophysics* 53(7): 1605–1610. doi:https://doi.org/10.3969/j.issn.0001-5733.2010.07.011.

Maercklin, N., G. Festa, S. Colombelli, and A. Zollo. 2012. Twin ruptures grew to build up the giant 2011 Tohoku, Japan, earthquake. *Scientific Reports* 2(709), doi: 10.1038/srep00709.

Marshall, P.D., and P.W. Basham. 1972. Discrimination between earthquakes and underground explosions employing an improved Ms scale. *Geophysical Journal of the Royal Astronomical Society* 28 (5): 431–458.

Meng, L., A. Inbal, and J.P. Ampuero. 2011. A window into the complexity of the dynamic rupture of the 2011 Mw 9 Tohoku-Oki earthquake. *Geophysical Research Letters* 38: L00G07.

Meng, L., J.-P. Ampuero, Y. Luo, et al. 2012a. Mitigating artifacts in back-projection source imaging with implications on frequency-dependent properties of the Tohoku-Oki earthquake. *Earth, Planets, and Space* 64(12): 1101–1109.

Meng, L., J.-P. Ampuero, A. Sladen, and H. Rendon. 2012b. High-resolution backprojection at regional distance: Application to the Haiti M7. 0 earthquake and comparisons with finite source studies. *Journal of Geophysical Research: Solid Earth (1978–2012)*. 117: B04313.

Meng, L., J.-P. Ampuero, J. Stock, et al. 2012c. Earthquake in a maze: compressional rupture branching during the 2012 Mw 8.6 Sumatra earthquake. *Science* 337 (6095): 724–726.

Olson, A.H., and R.J. Apsel. 1982. Finite faults and inverse theory with applications to the 1979 Imperial Valley earthquake. *Bulletin of the Seismological Society of America* 72 (6A): 1969–2001.

Reid, H.F. 1910. The California Earthquake of April 18, 1906: The Mechanics of the Earthquake/ by Harry Fielding Reid. Washington, D.C.: Carnegie Institute.

Reid, H.F. 1911. The elastic-rebound theory of earthquakes. *University of California Department of Geology Bulletin* 6(19): 413–444.

Rice, J.R. 1980. The mechanics of earthquake rupture, Physics of the Earth's Interior AM Dziewonski, E. Boschi, 555–649.

Ringdal, F., and E.S. Husebye. 1982. Application of arrays in the detection, location, and identification of seismic events. *Bulletin of the Seismological Society of America* 72 (6B): 201–224.

Rost, S., and C. Thomas. 2002. Array seismology: Methods and applications. *Reviews of Geophysics* 40 (3): 1008.

Roten, D., H. Miyake, and K. Koketsu. 2012. A rayleigh wave back-projection method applied to the 2011 Tohoku earthquake. *Geophysical Research Letters* 39: L02302.

Satriano, C., E. Kiraly, P. Bernard, and J.-P. Vilotte. 2012. The 2012 Mw 8.6 Sumatra earthquake: Evidence of westward sequential seismic ruptures associated to the reactivation of a N-S ocean fabric. *Geophysical Research Letters* 39: L15302.

Schweitzer, J., J. Fyen, and S. Mykkeltveit. 2002. Seismic arrays in Chap. 9. *New Manual of Seismological Observatory Practice* 1–51, doi: 10.2312/GFZ.NMSOP_r1_ch9

Spudich, P., and E. Cranswick. 1984. Direct observation of rupture propagation during the 1979 Imperial Valley earthquake using a short baseline accelerometer array. *Bulletin of the Seismological Society of America* 74 (6): 2083–2114.

Spudich, P., and L.N. Frazer. 1984. Use of ray theory to calculate high-frequency radiation from earthquake sources having spatially variable rupture velocity and stress drop. *Bulletin of the Seismological Society of America* 74 (6): 2061–2082.

Sufri, O., K.D. Koper, and T. Lay. 2012. Along-dip seismic radiation segmentation during the 2007 Mw 8.0 Pisco, Peru earthquake. *Geophysical Research Letters* 39: L08311.

Wagner, O.S., and C.A. Langston. 1989. Some pitfalls and trade-offs in source parameter determination using body wave modeling and inversion. *Tectonophysics* 166 (1): 101–114.

Walker, K.T., M. Ishii, and P.M. Shearer. 2005. Rupture details of the 28 March 2005 Sumatra Mw 8.6 earthquake imaged with teleseismic P waves. *Geophysical Research Letters* 32: L24303.

Walker, K.T., and P.M. Shearer. 2009. Illuminating the near-sonic rupture velocities of the intracontinental Kokoxili Mw 7.8 and Denali fault Mw 7.9 strike-slip earthquakes with global P wave back projection imaging. *Journal of Geophysical Research: Solid Earth (1978–2012)* 114: B02304.

Wang, D., and J. Mori. 2011. Frequency-dependent energy radiation and fault coupling for the 2010 Mw8. 8 Maule, Chile, and 2011 Mw9. 0 Tohoku, Japan, earthquakes. *Geophysical Research Letters* 38: L22308.

Wang, D., J. Mori, and T. Uchide. 2012. Supershear rupture on multiple faults for the Mw 8.6 Off Northern Sumatra, Indonesia earthquake of April 11, 2012. *Geophysical Research Letters* 39: L21307.

Xu, Y., K.D. Koper, O. Sufri, et al. 2009. Rupture imaging of the Mw 7.9 12 May 2008 Wenchuan earthquake from back projection of teleseismic P waves. *Geochemistry, Geophysics, Geosystems* 10 (4): 1–17.

Yagi, Y., A. Nakao, and A. Kasahara. 2012. Smooth and rapid slip near the Japan Trench during the 2011 Tohoku-oki earthquake revealed by a hybrid back-projection method. *Earth and Planetary Science Letters* 355: 94–101.

Yao, H., P. Gerstoft, P.M. Shearer, and C. Mecklenbrauker. 2011. Compressive sensing of the Tohoku-Oki Mw 9.0 earthquake: Frequency-dependent rupture modes. *Geophysical Research Letters* 38: L20310.

Yao, H., P.M. Shearer, and P. Gerstoft. 2012. Subevent location and rupture imaging using iterative backprojection for the 2011 Tohoku Mw 9.0 earthquake. *Geophysical Journal International* 190 (2): 1152–1168.

Yao, H., P.M. Shearer, and P. Gerstoft. 2013. Compressive sensing of frequency-dependent seismic radiation from subduction zone megathrust ruptures. *Proceedings of the National Academy of Sciences* 10 (12): 4512–4517.

Yue, H., T. Lay, and K.D. Koper. 2012. En echelon and orthogonal fault ruptures of the 11 April 2012 great intraplate earthquakes. *Nature* 490 (7419): 245–249.

Zhang, H., and Z. Ge. 2010. Tracking the rupture of the 2008 Wenchuan earthquake by using the relative back-projection method. *Bulletin of the Seismological Society of America* 100 (5B): 2551–2560.

Zhang, H., Z. Ge, and L. Ding. 2011. Three sub-events composing the 2011 off the Pacific coast of Tohoku Earthquake (Mw 9.0) inferred from rupture imaging by back- projecting teleseismic P waves. *Earth, Planets and Space* 63 (7): 595–598.

Zhang, H., J. Chen, and Z. Ge. 2012. Multi-fault rupture and successive triggering during the 2012 Mw 8.6 Sumatra offshore earthquake. *Geophysical Research Letters* 39: L22305.

Zhang, H., and Z. Ge. 2012. Rupture imaging of the 2010 Mw 8.8 Great Chilean earthquake using relative back-projection method. *Acta Scientiarum Naturalium Universitatis Pekinensis* 48 (4): 583–588.

Zhang, H., and Z. Ge. 2014. Rupture pattern of the Oct 23, 2011 Van-Merkez, Eastern Turkey earthquake. *Earthquake Science* 27 (3): 257–264.

Zhang, H., and Z. Ge. 2017. Stepover Rupture of the 2014 Mw 7.0 Yutian, Xinjiang, Earthquake. *Bulletin of the Seismological Society of America* 581–591.

Zhang, H., K.D. Koper, K. Pankow, and Z. Ge. 2017. Imaging the 2016 MW 7.8 Kaikoura, New Zealand earthquake with Teleseismic P Waves: A Cascading Rupture across Multiple Faults. *Geophysical Research Letters* 44. doi:https://doi.org/10.1002/2017GL073461.

Zhang, H., S. van der Lee, and Z. Ge. 2016. Multiarray rupture imaging of the devastating 2015 Gorkha, Nepal, earthquake sequence. *Geophysical Research Letters* 43(2), 584–591.

Chapter 2
Generalized Array Imaging on Rupture Processes of Earthquakes: Principle and Theoretical Tests

The rupture process of an earthquake is composed of nucleation of the earthquake at the hypocenter, subsequent rupture propagation on the fault, and stopping phase of the rupture. The rupture front usually proceeds at a speed less than the *P* wave velocity nearby the causative fault, accompanying with seismic energy releasing, thermal energy releasing, and strain energy consumed by the crack expansion. The ratio of the three kinds of energy depends on fault geometry, stress distribution, frictional property, media of the footwall and hanging wall, involvement of fluid, and tectonic stress. Among the three kinds of energy, only the seismic energy can be estimated with seismograms. The seismograms are convolution of the earthquake source, the structure of the earth's interior traveled through by the rays, and instrumental responses of seismometers. Typically, the instrumental responses are known. The structure of the earthquake from the earthquake to the seismometers can be expressed as Green's function. Given a velocity model like AK135 (Kennett et al. 1995), the Green's function can be calculated. Otherwise, if there are one-order less magnitude aftershocks with a similar focal mechanism to the mainshock, the waveforms of the aftershocks are suitable as empirical Green's function (EGF, Hartzell 1978). The slip models of the earthquakes can be obtained by fitting the observations on the basis of the known instrumental responses and Green's function using linear or nonlinear inversion algorithms. These slip models can provide very important references for post-earthquake rescuing and tsunami warning and show more rupture properties of the earthquakes. Thus, the investigation of the rupture processes of earthquakes plays a significant role in anti-seismic hazard and studying earthquake sources.

Rupture models of earthquakes can be derived by inverting the seismic body waves or surface waves. To stabilize the inversion and make computation effective, several constraints such as smoothing, minimum seismic moment, and non-reverse slip are placed during the inversion. Moreover, the solution of the inversion strongly depends on the initial model. Additionally, the velocity model used in the calculation of the Green's function deviates greatly from the structure of the earth; this leads to systematic or subjective errors in the resultant rupture model.

H. Zhang, *Imaging the Rupture Processes of Earthquakes Using the Relative Back-Projection Method*, Springer Theses,
https://doi.org/10.1007/978-3-662-55239-1_2

Besides the seismic data, other kinds of data such as water height of tsunami and displacement of the surface after an earthquake surveyed by GPS or Interferometric Satellite Aperture Radar (InSAR) can be incorporated into the inversion of the rupture model. This can greatly improve the accuracy of the rupture model since multiple datasets provide much more tight constraints for the inversion. Nevertheless, since the global seismic data can be fetched shortly after an earthquake, most of the rupture models are derived from the seismic data.

Also, there is another way, namely generalized array techniques such as back-projection method (BPM, Ishii et al. 2005), to quickly obtain the rupture processes of the earthquakes. The back-projection method can be applied to regional arrays and first used by Ishii et al. (2005) to study the spatial-temporal rupture of the December 26, 2004, Mw 9.2 Sumatra–Andaman earthquake. Actually, the rupture front is tracked by the BPM since the high-frequency seismic waveforms radiated from the rupture front (Madariaga 1977, 1983) are used by the BPM. Because the locations and rupture times of the rupture front can be traced, the rupture dimensions as well as rupture velocity can be accurately estimated.

Compared with the rupture models derived by the inversions, the back-projection imaging has advantages as follows. First, the higher-frequency data were used. Thus, the back-projection imaging can provide more rupture details in theory. Second, less known information is required to carry out the imaging procedure. Third, relative back-projection imaging depends less on the velocity model, since all the subevents are imaged relative to the epicenter. Given the advantages listed above and the easy operation, the back-projection method has been widely used to study the rupture processes of large earthquakes such as the 2004 Mw 9.2 Sumatra–Andaman earthquake (Ishii et al. 2005, 2007; Kruger and Ohrnberger, 2005; Du et al. 2009; Liu et al. 2007), the 2010 Mw 8.8 Chilean earthquake (Kiser and Ishii, 2011; Wang and Mori 2011; Liu et al. 2010), the 2011 Mw 9.0 Tohoku, Japan, earthquake (Ishii 2011; Kiser and Ishii 2012; Koper et al. 2011a, b; Maercklin et al. 2012; Meng et al. 2012a; Wang and Mori 2011; Roten et al. 2012; Yagi et al. 2012; Yao et al. 2011, 2012; Zhang et al. 2011), and the 2012 Mw 8.6 Sumatra offshore earthquake (Meng et al. 2012c; Satriano et al. 2012; Yue et al. 2012; Zhang et al. 2012). The researches on these large earthquakes dramatically renew one's conventional understanding on earthquake sources, such as frequency-dependent rupture, segmentation, and multiple-fault rupture, which will be discussed in the following chapters.

The advantages of the BPM make it complementary to the finite fault modeling in the rupture of earthquakes. For the inversion of the slip model, rupture velocity is strongly coupled with slip rates of fault patches (Spudich and Frazer 1984). To address this issue, the rupture velocity can be accurately estimated by the BPM. Thus, providing the rupture velocity derived by the BPM can lead to a better estimate to the slip models of the earthquakes using finite fault modeling.

Although the back-projection method has been prevalently applied to image the rupture processes of the earthquakes, there still exist several issues in the back-projection imaging. The most prominent issue is the imaging artifacts, which could lead to an interpretation of fake rupture. The artifacts mainly have two sources: One is the subevents releasing more high-frequency energy, which results in imaging "swimming," and the other one is the multiples or depth phases from earlier subevents. To suppress the first kind of artifacts, a relative back-projection method (RBPM) is proposed in this book. For the second kind of artifacts, a combination of back-projection imaging from multiple arrays can be used to mitigate this kind of artifacts (Zhang et al. 2016a, 2017; Zhang and Ge 2016a, b).

To demonstrate the effectiveness of the RBPM to suppress "swimming" artifacts, the principle of the beamforming used in the back-projection imaging is first deduced. Next, five of the most used generalized array techniques are introduced. More importantly, the procedure of the RBPM is elaborated. Finally, using the synthesized waveforms, the temporal and spatial resolutions of the RBPM are explored and the imaging results derived by the RBPM are compared with those from the traditional BPM.

2.1 Principle of Beamforming

Large earthquakes have a duration of 100 s up to 500 s and a rupture length of several hundred kilometers up to 1300 km (Ammon et al. 2005; Lay et al. 2005; Ishii et al. 2005). The rupture process of a large earthquake can be simulated by a series of subevents rupturing in distinct locations at different rupture times (Ge and Chen 2008). All the subevents can be tracked by the back-projection method, and the rupture process of the earthquake is thus imaged. The key technique used in the back-projection method is slant stacking. To demonstrate the effectiveness of the BPM, the principle of the slant stacking is elaborated as follows.

Assuming that a large earthquake occurred, the seismic waves from the earthquake are recorded by an array composed of M stations $\{\vec{x}_k, k = 1, \ldots, M\}$ as shown in Fig. 2.1. According to the displacement representation theorem by Aki

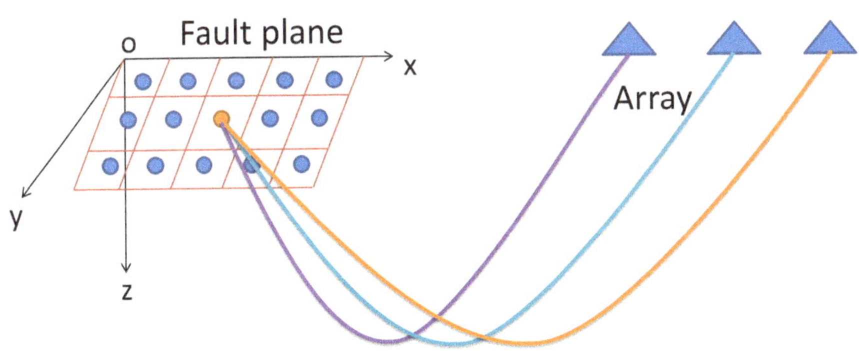

Fig. 2.1 Discretization of a fault plane. When some subfault (orange dot) ruptures, the radiated waves are recorded by an array composed of multiple seismometers

and Richards (2002), dislocation on the fault plane $\sum$ generates the displacement recorded at the kth station $\vec{x}_k$ as

$$u_n(\vec{x}_k, t) = \int_{-\infty}^{+\infty} d\tau \iint_{\sum} \left[u_i\left(\vec{\xi}, \tau\right)\right] c_{ijpq} \upsilon_j \frac{\partial}{\partial \xi_q} G_{np}\left(\vec{x}_k, t-\tau; \vec{\xi}, 0\right) d\sum, \tag{2.1}$$

where $u_n(\vec{x}_k, t)$ is the displacement along the n direction recorded at the station $\vec{x}_k$, $u_i\left(\vec{\xi}, \tau\right)$ is dislocation at $\vec{\xi}$ along the direction j, c_{ijpq} is the fourth-order elastic tensor, υ_j is the projection to the direction j of the unit normal vector of the fault, and $G_{np}\left(\vec{x}_k, t-\tau; \vec{\xi}, 0\right)$ is the Green's function, which is the displacement along the direction n at the station $\vec{x}_k$ at the moment $t - \tau$ generated by the point force exerting at $\vec{\xi}$ on the fault along the direction p at 0 s.

Provided that Green's function is independent of time, the representation theorem can be expressed as

$$u_n(\vec{x}_k, t) = \int_{-\infty}^{+\infty} d\tau \iint_{\sum} \left[u_i\left(\vec{\xi}, \tau\right)\right] c_{ijpq} \upsilon_j \frac{\partial}{\partial \xi_q} G_{np}\left(\vec{x}_k, t; \vec{\xi}, \tau\right) d\sum, \tag{2.2}$$

where Green's function $G_{np}\left(\vec{x}_k, t; \vec{\xi}, \tau\right)$ is the displacement along the direction n at the station $\vec{x}_k$ at the moment t generated by the point force exerting at $\vec{\xi}$ on the fault along the direction p at the moment τ.

There are two end-members to describe heterogeneous dynamic rupture on the fault. One is the barrier model proposed by Das and Aki (1977), Aki (1984), and the other one is the asperity model proposed by Lay and Kanamori (1981), Rudnicki and Kanamori (1981). Both models show that the faults ruptured by earthquakes are not totally a flat plane, and elastic coefficients as well as fault strength are unevenly distributed on the fault. Thus, elastic coefficient tensor and unit normal vector of the fault plane are related to $\vec{\xi}$ as $c_{ijpq}\left(\vec{\xi}\right)$ and $\upsilon_j\left(\vec{\xi}\right)$. After substituting the moment density tensor $m_{pq}\left(\vec{\xi}, t\right) = u_i\left(\vec{\xi}, \tau\right) c_{ijpq} \upsilon_j$ (Aki and Richards 2002), the displacement at the station $\vec{x}_k$ can be expressed as

$$u_n(\vec{x}_k, t) = \iint_{\sum} m_{pq}\left(\vec{\xi}, t\right) * G_{np,q}\left(\vec{x}_k, t; \vec{\xi}\right) d\sum, \tag{2.3}$$

where $G_{np,q}\left(\vec{x}_k, t; \vec{\xi}\right) = \frac{\partial}{\partial \xi_q} G_{np}\left(\vec{x}_k, t; \vec{\xi}\right)$.

The data used in back-projection are velocity seismograms, which have more high-frequency component relative to displacement. The high-frequency component of the seismic data is what be used in the back-projection. The velocity record at the station $\vec{x}_k$ is a derivative of displacement as

$$v_n(\vec{x}_k, t) = \iint_{\sum} \frac{\mathrm{d}}{\mathrm{d}t}\{m_{pq}\left(\vec{\xi}, t\right) * G_{np,q}\left(\vec{x}_k, t; \vec{\xi}\right)\}d\sum, \tag{2.4}$$

where $v_n(\vec{x}_k, t) = \dot{u}_n(\vec{x}_k, t)$.

Since the convolution has a differentiating property (Xu et al. 2005) as

$$\frac{\mathrm{d}}{\mathrm{d}t}[f_1(t) * f_2(t)] = \frac{\mathrm{d}f_1(t)}{\mathrm{d}t} * f_2(t),$$

where $f_1(t)$ and $f_2(t)$ are as a function of time, the velocity can be rewritten as

$$v_n(\vec{x}_k, t) = \iint_{\sum} \dot{m}_{pq}\left(\vec{\xi}, t\right) * G_{np,q}\left(\vec{x}_k, t; \vec{\xi}\right)d\sum, \tag{2.5}$$

where $\dot{m}_{pq}\left(\vec{\xi}, t\right) = \frac{\mathrm{d}}{\mathrm{d}t}\{m_{pq}\left(\vec{\xi}, t\right)\}$.

The rupture of large earthquakes can be simulated by slip evolving with time on a series of subfaults (Ge and Chen 2008). The discretization of the fault plane into multiple subfaults is the typical strategy of the seismic inversion for the rupture models (Kikuchi and Kanamori 1982; Olsen and Apsel 1982; Hartzell and Heaton 1983). Similarly, the back-projection is also taking the same strategy to process the source region. Thus, the fault plane $\sum$ is discretized into N subfaults with an area of ΔA. Thus, the velocity at the station $\vec{x}_k$ can be expressed as

$$v_n(\vec{x}_k, t) = \sum_{i=1}^{N} \dot{m}_{pq}\left(\vec{\xi}_i, t\right) * G_{np,q}\left(\vec{x}_k, t; \vec{\xi}_i\right)\Delta A. \tag{2.6}$$

The moment tensor of the ith subfault is expressed as

$$M_{pq}\left(\vec{\xi}_i, t\right) = m_{pq}\left(\vec{\xi}_i, t\right)\Delta A.$$

The expression of the velocity can be changed as

$$v_n(\vec{x}_k, t) = \sum_{i=1}^{N} \dot{M}_{pq}\left(\vec{\xi}_i, t\right) * G_{np,q}\left(\vec{x}_k, t; \vec{\xi}_i\right). \tag{2.7}$$

Another seismic approach to study discontinuities in the earth's interior is teleseismic P wave receiver function. For example, the P wave receiver function has been used to clearly resolve the crustal structure of the North America Mid-Continent Rift (Zhang et al. 2016b). The principle of the receiver function indicates the teleseismic direct P wave can be thought as an impulse source for its multiples and the Green's function can be derived by deconvolving the radial component of the waveforms with the vertical component (Langston 1976); this indicates that the Green's function of the direct P waves can be expressed with Dirac-δ function. The spatial derivative of the Green's function can be decomposed to be summation of direct P waves and multiples (Yao et al. 2012) as

$$G_{np,q}\left(\vec{x}_k, t; \vec{\xi}_i\right) = G^{d}_{np,q}\left(\vec{x}_k; \vec{\xi}_i\right)\delta(t - t_{ik}) + G^{m}_{np,q}\left(\vec{x}_k, t; \vec{\xi}_i\right),$$

where $G^{d}_{np,q}\left(\vec{x}_k, t; \vec{\xi}_i\right)$ and $G^{m}_{np,q}\left(\vec{x}_k, t; \vec{\xi}_i\right)$ are the spatial derivatives of the direct P waves and multiples, respectively, and t_{ik} is the travel time from the source $\vec{\xi}_i$ to the station $\vec{x}_k$.

Replacing the Green's function in Eq. 2.7 with the above equation, the velocity can be expressed as

$$v_n(\vec{x}_k, t) = \sum_{i=1}^{N}\left\{G^{d}_{np,q}\left(\vec{x}_k; \vec{\xi}_i\right)\dot{M}_{pq}\left(\vec{\xi}_i, t - t_{ik}\right) + v^{m}_{n}(\vec{x}_k, t)\right\}, \tag{2.8}$$

where $v^{m}_{n}(\vec{x}_k, t)$ is the waveforms without the direct P waves.

Besides the multiples, there are background or man-made noises in the waveforms. This means that noises need to be included in the waveforms as

$$v^{m}_{n}(\vec{x}_k, t) = \sum_{i=1}^{N}\{G^{m}_{np,q}\left(\vec{x}_k, t; \vec{\xi}_i\right) * \dot{M}_{pq}\left(\vec{\xi}_i, t\right)\} + n(\vec{x}_k, t),$$

where $n(\vec{x}_k, t)$ is the noise at the kth station. Thus, even if incorporating the noise term, the expession of the velocity still stands.

The focal mechanism on the subfault is constant when the dimensions of the subfault are small enough. Under this assumption, the derivative of moment tensor relative to time can be expressed as

$$\dot{M}_{pq}\left(\vec{\xi}_i, t\right) = M_{pq}\left(\vec{\xi}_i\right)\dot{s}\left(\vec{\xi}_i, t\right), \tag{2.9}$$

where $\dot{s}\left(\vec{\xi}_i, t\right)$ is the rate of the source time function at the subfault $\vec{\xi}_i$.

Chen and Gu (2008) provide an expression of the source time function as

$$s(t) = \begin{cases} 0, t < 0, \\ 1, t \to \infty. \end{cases}$$

For a subfault $\vec{\xi}_i$, the rise time of the rupture T_s^i is limited. Different subfaults have distinct initial rupture time t_i^0. Without loss of generality, the source function can be set to a ramp function as

$$s\left(\vec{\xi}_i, t\right) = \begin{cases} 0, & t < t_i^0, \\ \frac{t}{T_s^i}, & t_i^0 \le t \le t_i^0 + T_s^i, \\ 1, & t > t_i^0 + T_s^i. \end{cases}$$

The derivative of the source time function with respect to time is derived as

$$\dot{s}\left(\vec{\xi}_i, t\right) = \begin{cases} 0, & t < t_i^0, \\ \frac{1}{T_s^i}, & t_i^0 \le t \le t_i^0 + T_s^i, \\ 0, & t > t_i^0 + T_s^i. \end{cases}$$

Substituting the Heaviside function into the rate of source time function, the above equation can be changed to

$$\dot{s}\left(\vec{\xi}_i, t\right) = \frac{H\left(t - t_i^0\right) H\left(-t + t_i^0 + T_s^i\right)}{T_s^i}, \quad (2.10) \tag{2.10}$$

where $H(t)$ is Heaviside function as

$$H(t) = \begin{cases} 0, t < 0, \\ 1, t \ge 0. \end{cases}$$

Substituting the Eq. 2.10 into the Eq. 2.9, the time derivative of the moment tensor can be expressed as

$$\dot{M}_{pq}\left(\vec{\xi}_i, t\right) = \frac{M_{pq}\left(\vec{\xi}_i\right) H\left(t - t_i^0\right) H\left(-t + t_i^0 + T_s^i\right)}{T_s^i}. \tag{2.11}$$

After incorporating the Eq. 2.11 into the Eq. 2.8, the velocity at the station $\vec{x}_k$ can be deduced to

$$v_n(\vec{x}_k, t) = \sum_{i=1}^{N} \left\{ A(\vec{x}_k, \vec{\xi}_i) H\left(t - t_{ik} - t_i^0\right) H\left(-t + t_{ik} + t_i^0 + T_s^i\right) \right\} + v_n^m(\vec{x}_k, t), \tag{2.12}$$

where $A\left(\vec{x}_k, \vec{\xi}_i\right) = \frac{G_{np,q}^d\left(\vec{x}_k; \vec{\xi}_i\right) M_{pq}\left(\vec{\xi}_i\right)}{T_s^i}$, which is the amplitude of the direct P waves.

Before conducting beamforming or slant stacking, the velocity record for each station is normalized with the maximum amplitude $A_{\max}(\vec{x}_k)$. The normalized velocity records can be expressed as

$$\bar{v}_n(\vec{x}_k, t) = \sum_{i=1}^{N} \{\bar{A}(\vec{x}_k, \vec{\xi}_i) H(t - t_{ik} - t_i^0) H(-t + t_{ik} + t_i^0 + T_s^i)\} + \bar{v}_n^m(\vec{x}_k, t), \quad (2.13)$$

where $\bar{v}_n(\vec{x}_k, t) = v_n(\vec{x}_k, t)/A_{\max}(\vec{x}_k)$, $\bar{A}\left(\vec{x}_k, \vec{\xi}_i\right) = A\left(\vec{x}_k, \vec{\xi}_i\right)/A_{\max}(\vec{x}_k)$, and $\bar{v}_n^m(\vec{x}_k, t) = v_n^m(\vec{x}_k, t)/A_{\max}(\vec{x}_k)$. For the subfault l, the seismograms at M stations are back-projected and stacked up. The beam thus is formed as

$$B\left(\vec{\xi}_l, t\right) = \frac{1}{M} \sum_{k=1}^{M} \sum_{i=1}^{N} \{\bar{A}(\vec{x}_k, \vec{\xi}_i) H(t + t_{lk} - t_{ik} - t_i^0) H(-t - t_{lk} + t_{ik} + t_i^0 + T_s^i)\} + \frac{1}{M} \sum_{k=1}^{M} \bar{v}_n^m(\vec{x}_k, t + t_{lk}), \quad (2.14)$$

where t_{lk} is the travel time from the l th subfault to the k th station, and $B\left(\vec{\xi}_l, t\right)$ is the beam back-projected at the l th subfault $\vec{\xi}_l$.

The difference between the travel times of the direct P waves from the ith subfault $\vec{\xi}_l$ and the ith subfault $\vec{\xi}_i$ to the station $\vec{x}_k$ is distinct. Back-projection is to reverse the time of the waveforms back to the initial rupture time of the subfault $\vec{\xi}_l$. The waveforms at different stations would be completely coherent when the l th subfault $\vec{\xi}_l$ is just the i th subfault $\vec{\xi}_i$. This back-projection operation significantly suppresses the beam of the multiples; this leads to negligibility of the second term of the Eq. 2.14. In this case, the beam of the velocity records can be expressed as

$$\begin{aligned}
B\left(\vec{\xi}_l, t\right) &= \frac{1}{M} \sum_{k=1}^{M} \sum_{i=1}^{N} \{\bar{A}(\vec{x}_k, \vec{\xi}_i) H(t + t_{lk} - t_{ik} - t_i^0) H(-t - t_{lk} + t_{ik} + t_i^0 + T_s^i)\} \delta_{il}, \\
&= \left\{\frac{1}{M} \sum_{k=1}^{M} \bar{A}(\vec{x}_k, \vec{\xi}_l)\right\} H(t - t_l^0) H(-t + t_l^0 + T_s^l), \\
&= \left\{\frac{1}{M} \sum_{k=1}^{M} \frac{G_{np,q}^d\left(\vec{x}_k; \vec{\xi}_l\right) M_{pq}\left(\vec{\xi}_l\right)}{T_s^l A_{\max}(\vec{x}_k)}\right\} H(t - t_l^0) H(-t + t_l^0 + T_s^l), \qquad (2.15) \\
&= \left\{\frac{1}{M} \sum_{k=1}^{M} \frac{G_{np,q}^d\left(\vec{x}_k; \vec{\xi}_l\right) M_{pq}\left(\vec{\xi}_l\right)}{A_{\max}(\vec{x}_k)}\right\} \frac{H(t - t_l^0) H(-t + t_l^0 + T_s^l)}{T_s^l}, \\
&= \bar{A}^d\left(\vec{\xi}_l\right) \dot{s}\left(\vec{\xi}_l, t\right),
\end{aligned}$$

where δ_{il} is Kronecker delta function, and $\bar{A}^d\left(\vec{\xi}_l\right)$ is a scale factor associated with the generalized array at the l th subfault $\vec{\xi}_l$. This proofs that the beam of the waveforms is the product of the derivative of the source time function and the array scale factor. Thus, the subfault can be effectively tracked by the back-projection method, which has been proofed above in theory.

2.2 Typical Generalized Array Approaches

The theoretical proof of the beamforming described above confirms the reliability of these array techniques as beamforming is fundamental to the generalized array techniques. Most of the data used by the generalized array techniques are the direct *P* waves because the *P* waves arrive earlier at some station. The earlier arrivals facilitate picking-up of the arrival time of the first *P* waves; this is significantly important for calibrating the travel times through one-dimensional velocity model to the heterogeneous three-dimensional earth's interior.

Generalized array techniques can be classified into several categories on the basis of the wave phases or the domain in which the techniques are exploited. From the view of wave phase, teleseismic direct *P* waves were first used to study the 2004 Mw 9.2 Sumatra–Andaman earthquake (Ishii et al. 2005; Krüger and Ohrnberger 2005) and were the most kind of data used until now (Walter et al. 2005; Xu et al. 2009; Zhang and Ge 2010; Kiser and Ishii 2011; Koper et al. 2011a, b; Wang and Mori 2011; Meng et al. 2011; Yao et al. 2012; Zhang and Ge 2014). Besides the direct *P* waves, the direct *S* or *Pn* waves were used to investigate the local or regional earthquakes (Allmann and Shearer 2007; Meng et al. 2012b). Additionally, the fundamental surface waves were employed to image the rupture process of the earthquakes (Yue et al. 2012; Roten et al. 2012), though the resolutions of surface waves are less than those of the body waves. Meanwhile, the generalized array techniques can also be grouped into time-domain type (Ishii al el. 2005; Walter et al. 2005; Xu et al. 2009; Zhang and Ge 2010; Roten et al. 2012; Yagi et al. 2012) and frequency-domain type (Goldstein and Archuleta 1991; Meng et al. 2011; Yao et al. 2011).

Each kind of generalized array techniques has their own advantages. Five techniques including traditional back-projection method, cross-correlation back-projection, surface-wave back-projection method, multiple signal classification method, and compression sensing will be introduced as follows.

2.2.1 *Traditional Back-Projection Method*

Ishii et al. (2005) were the first ones to apply the traditional back-projection method to study the rupture process of the 2004 Mw 9.2 Sumatra–Andaman earthquake using teleseismic *P* waves, though the array techniques had been used in the research of rupture of local and regional earthquakes (Bolt et al. 1982; Spudich and Cranswick 1984; Goldstein and Archuleta 1991; Huang 2001). The traditional back-projection method (TBPM) is based on beamforming, but with a significant improvement to eliminate the assumption of plane wave impingement, this leads to usage of the TBPM for large teleseismic events instead of local or regional earthquakes. The procedure of the TBPM was elaborated by Ishii et al. (2007) and will be briefly introduced below.

At first, the source region of an earthquake is gridded into N potential subevents. The travel time of P waves from a potential subevent to some station $\{\vec{x}_k, k = 1, \ldots, M\}$ can be calculated with a one-dimensional velocity model such as AK135 (Kennett et al. 1995). However, the real earth's interior is heterogeneous. A time correction term has to be added to compensate this deviation of earth's structure. The first P arrivals generated by the subevent at the hypocenter can be accurately picked up. Thus, taken the hypocenter as a reference, the time calibration at the station $\vec{x}_k$ can be estimated as

$$\delta t_k = t_k^o - t_{ek}^c,$$

where t_k^o is the observed P travel time of the hypocenter to the station $\vec{x}_k$, and t_{ek}^c is the travel time of the hypocenter to the station $\vec{x}_k$ calculated with a given velocity model. For each potential subevent, the travel times to the stations can be calculated. After the time calibration being incorporated, the direct P waves are back-projected by converting the recording time to the rupture time. For each rupture time, the waveforms are summed up to obtain the beams. Thus, the beam generated by the ith potential subevent can be expressed as

$$B\left(\vec{\xi}_i, t\right) = \frac{1}{M}\sum_{k=1}^{M} \alpha_k v\left(\vec{x}_k, t + t_{ik}^c + \delta t_k\right), \tag{2.16}$$

where $\alpha_k = \frac{p_k}{A_k^{\max}} w_k$, p_k is the polarity of the P waves at the station $\vec{x}_k$, $A_k^{\max}$ is the maximum absolute amplitude of the kth waveform, and w_k is the weight of the kth waveform such as signal-to-noise ratio (SNR), $v(\vec{x}_k, t)$ is the velocity at the station $\vec{x}_k$, and t_{ik}^c is the calculated travel time of the ith potential subevent to the station $\vec{x}_k$.

It is inappropriate to only use the initial time calibration to compensate the one-dimensional velocity model to the real earth, when the studied earthquakes have very large rupture dimensions such as the 2004 Mw 9.2 Sumatra-Andaman earthquake (Ammon et al. 2005; Ishii et al. 2005; Lay et al. 2005). Relative to the large mainshock, the aftershocks have much shorter durations and may have sharper first arrivals. Since the aftershocks are typically distributed all over the fault of the mainshock, the arrivals of the aftershocks can be used as the time calibrations when there are enough aftershocks. Thus, given L aftershocks, the time calibrations of the ith potential subevent to the station $\vec{x}_k$ (Ishii et al. 2007) can be estimated as

$$\delta t_{ik} = \frac{\sum_{j=1}^{L}\left\{w_j \delta t_{jk}/\Delta_{ij}\right\}}{\sum_{j=1}^{L}\left\{w_j/\Delta_{ij}\right\}},$$

where w_j is the weight of the jth aftershock, Δ_{ij} is the distance of the jth aftershock to the ith potential subevent, and $\delta t_{jk} = t_{jk}^o - t_{jk}^c$ is the time calibration of the jth aftershock to the station $\vec{x}_k$, t_{jk}^o and t_{jk}^c are the observed and calculated travel times of the direct P waves generated from the jth aftershock to the station $\vec{x}_k$, respectively.

Using the more accurate time calibrations, the beam at the *i*th potential subevent is formed as

$$B\left(\vec{\xi}_i, t\right) = \frac{1}{M}\sum_{k=1}^{M} \alpha_k v\left(\vec{x}_k, t + t_{ik}^c + \delta t_{ik}\right) = \frac{1}{M}\sum_{k=1}^{M} \bar{v}_{ik}(t), \tag{2.17}$$

where $\bar{v}_{ik}(t)$ is weighted, normalized velocity record at the station $\vec{x}_k$.

To reduce the interference of coda of the earlier-ruptured subevents, the *N*th root stacking technique (Muirhead 1968; Kanasewih et al. 1973) was applied by Xu et al. (2009). The *N*th root stacking can be expressed as

$$B\left(\vec{\xi}_i, t\right) = \left|B\prime\left(\vec{\xi}_i, t\right)\right|^{1/N} \cdot \mathrm{sign}\left\{B'\left(\vec{\xi}_i, t\right)\right\}, \quad (2.18) \tag{2.18}$$

where $B'\left(\vec{\xi}_i, t\right) = \frac{1}{M}\sum_{k=1}^{M} |\bar{v}_{ik}(t)|^{1/N} \cdot \mathrm{sign}\{\bar{v}_{ik}(t)\}$, $\mathrm{sign}\{\cdot\}$ is sign function, and N is an integer. When $N = 1$, Eq. 2.18 is the same as Eq. 2.17; this indicates the linear slant stacking is a special case of the *N*th root stacking.

A large earthquake can be thought as a series of subevents (Ge and Chen 2008). The power of the beam at the *i*th potential subevent in the lth time window with a length of T_w can be expressed as

$$P_{il} = \frac{1}{T_w}\int_{T_l}^{T_l + T_w} \left|B\left(\vec{\xi}_i, t\right)\right|^2 \mathrm{d}t, \tag{2.19}$$

where $T_l = (l - 1)T_s$, and T_s is the step of the moving time window. For each step, the potential subevent with the maximum power is identified as the real subevent. Thus, the rupture front evolving with rupture time is obtained.

2.2.2 *Cross-Correlation Back-Projection Method*

Cross-correlation is prevalent in signal processing such as communication, industrial seismology, and signal recognition and relocation (Shear 1997). This technique can be understood in both time and frequency domains (Anstey 1964). In time domain, the cross-correlation function indicates the similarity of one waveform to another one. In frequency time, the cross-correlation function of one waveform is equivalent to a filter with the same amplitude spectra, but the opposite phase spectra to the reference waveform.

Flether et al. (2006) are the first ones to apply the cross-correlation back-projection method to image the rupture propagation of the 2004 Parkfield earthquake using a local, small-aperture array nearby the earthquake. For the small-aperture array, the impinging waves can be thought to be plane waves (Schweitzer et al. 2002). The delay time of the plane wave impinging into the array between two stations in the array can be expressed as

$$\tau_{ij} = \vec{s} \cdot \vec{r}_{ij} + \delta t_i - \delta t_j, \tag{2.20}$$

where $\vec{s} = (s_E, s_N, s_z)$ is slowness vector in a Cartesian coordinate, $||s|| = 1/c$, c is velocity of the phase used, $\vec{r}_{ij}$ is the distance vector between stations i and j, and δt_i and δt_j are the travel time calibrations of the stations i and j. The cross-correlation between the waveforms at the two stations can be written as

$$cc_{ij} = \frac{\sum_t [v(\vec{x}_i, t) - \bar{v}(\vec{x}_i)]\left[v(\vec{x}_j, t) - \bar{v}(\vec{x}_j)\right]}{\sqrt{\sum_t [v(\vec{x}_i, t) - \bar{v}(\vec{x}_i)]^2 \left[v(\vec{x}_j, t) - \bar{v}(\vec{x}_j)\right]^2}}, \quad (2.21) \qquad (2.21)$$

where $\bar{v}(\vec{x}_i)$ and $\bar{v}(\vec{x}_j)$ are the means of the waveforms at the stations i and j, respectively.

For a horizontal slowness vector (s_E^p, s_N^q), the vertical slowness can be obtained as

$$s_Z = \sqrt{\frac{1}{c^2} - s_E^p \cdot s_E^p - s_N^q \cdot s_N^q}.$$

For each station, the delay time can be estimated with the Eq. 2.20 and the cross-correlation coefficient can be also derived with the Eq. 2.21. The cross-correlation coefficient of the whole array can be obtained by averaging those of all the stations as

$$\overline{cc}_{pq} = \frac{1}{M(M-1)} \sum_{k=1}^{M} \sum_{k=1}^{M} cc_{ij}(s_E^p, s_N^q),\ i \neq j. \tag{2.22}$$

At each time step, the plane wave with the horizontal slowness makes the array cross-correlation coefficient maximum stem from the real subevent of the earthquake. According to the azimuth of the subevent and apparent velocity of the plane wave, the location of the subevent can be obtained. Thus, the subevents of the earthquake evolving with time can be derived.

Besides the cross-correlation between the waveforms at different stations, the theoretical Green's function can also be used to calculate the cross-correlation coefficient with the waveforms at the stations (Yagi et al. 2012). This technique can eliminate the interference of depth phases (*pP* and *sP*) if the Green's function is correct. However, it is really hard to obtain correct high-frequency Green's function because it requires a high-resolution velocity model.

For the Green's function, the cross-correlation coefficient with the observations can be expressed as

$$cc\left(\vec{x}_i, \vec{\xi}_j, t\right) = \frac{\int_{t_w} v(\vec{x}_i, \tau + t) g(\vec{x}_j, \vec{\xi}_i, \tau) d\tau}{V_i \sqrt{\int_{t_w} g^2(\vec{x}_j, \vec{\xi}_i, \tau) d\tau}}, \tag{2.23}$$

where $V_i = \max\left(\sqrt{\int_w v^2(\vec{x}_i, \tau + t)}, 0 \le t \le t_w\right)$, t_w is the time length of the waveforms, and $g(\vec{x}_j, \vec{\xi}_i, \tau)$ is Green's function. The beam of the cross-correction functions can be derived as

$$B\left(\vec{\xi}_i, t\right) = |B'\left(\vec{\xi}_i, t\right)|^{1/N} \cdot \text{sign}\left\{B'\left(\vec{\xi}_i, t\right)\right\}, \tag{2.24}$$

where $B'\left(\vec{\xi}_i, t\right) = \frac{1}{M}\sum_{k=1}^{M} |cc\left(\vec{x}_i, \vec{\xi}_j, t\right)|^{1/N} \cdot \text{sign}\left\{cc\left(\vec{x}_i, \vec{\xi}_j, t\right)\right\}$. The power of the beam can be expressed the same as the Eq. 2.31. Similarly, the subevent at each time step is identified with the maximum power. Thus, the rupture of the earthquake proceeding with time can be imaged.

2.2.3 *Multiple Signal Classification*

Multiple signal classification (MUSIC) was proposed by Schmidt (1986) to discriminate the sources with different direction of arrivals (DOA). Goldstein and Archuleta (1991) applied this approach in frequency-wavenumber domains to study the earthquake sources using a local, small-aperture array. As the number of global stations increases, the MUSIC has been used to study the rupture processes of large earthquakes (Meng et al. 2011, 2012c).

The imaging procedure of the MUSIC is demonstrated as follows. First of all, the waveforms (M) in the first time window (with a length of t_w) are aligned to the first direct P arrivals in a manually pick way or using a multiple-channel cross-correlation approach (Vandecar and Crosson 1990). The covariance matrix of the M waveforms in the lth time window can be expressed as

$$R_{pq}^l = \left\langle v_l(\vec{x}_p, t) v_l^\dagger(\vec{x}_p, t)\right\rangle_t,$$

where $\langle\cdot\rangle_t$ indicates an operator averaging time and $\dagger$ is the Hermitian conjugate operator. Next, the covariance matrix $\boldsymbol{R}^l$ will be solved to obtain the eigenvectors.

It is reasonable to assume that the $\boldsymbol{R}^l$ has Q eigenvectors ($Q < M$) and λ_q is the qth eigenvector. If the wave field is decomposed into plane waves, the eigenvectors of the seismic covariance matrix $\boldsymbol{R}^l$ become the propagation vectors of the plane waves (Goldstein and Archuleta 1987). Thus, with respect to the hypocenter, the plane waves generated by the qth subevent have a phase shift as

$$\boldsymbol{u}\left(\hat{\xi}_q\right) = \left[e^{i\omega\left(t_{q1}-t_{e1}\right)}, \ldots, e^{i\omega\left(t_{qM}-t_{eM}\right)}\right]^T,$$

where ω is circular frequency, T is transpose of a matrix, and t_{qk} and t_{ek} are travel times of the kth station to the hypocenter ξ_e and the qth subevent $\hat{\xi}_q$.

Assuming that the background noise is a random variable normally distributed with a variance of σ^2, the seismic covariance matrix $\boldsymbol{R}^l$ can be decomposed as

$$\boldsymbol{R}^l = \boldsymbol{U}\boldsymbol{S}\boldsymbol{U}^\dagger + \sigma^2\boldsymbol{I},$$

where $\boldsymbol{U} = (\boldsymbol{u}(\hat{\xi}_1)^T \ldots \boldsymbol{u}\left(\hat{\xi}_Q\right)^T 0 \ldots 0\Big)$ is a $M \times M$ matrix, $\boldsymbol{I}$ is a unit matrix,

$$\boldsymbol{S} = \begin{pmatrix} \lambda_1^2 & 0 & \cdots & \cdots & \cdots & 0 \\ 0 & \ddots & \cdots & \cdots & \cdots & \vdots \\ \vdots & \vdots & \lambda_Q^2 & \cdots & \cdots & \vdots \\ \vdots & \vdots & \vdots & 0 & \cdots & \vdots \\ \vdots & \vdots & \vdots & \vdots & \ddots & \vdots \\ 0 & \cdots & \cdots & \cdots & \cdots & 0 \end{pmatrix}.$$

The Q eigenvectors are composed of the signal subspace. The left $M - Q$ eigenvectors represent the noise subspace. The signal eigenvectors are orthogonal to the noise eigenvectors as

$$\boldsymbol{e}_p \cdot \boldsymbol{u}\left(\hat{\xi}_q\right) = 0, p = Q+1, \ldots, M; q = 1, \ldots, Q.$$

The seismic covariance matrix $\boldsymbol{R}^l$ can thus be decomposed to the summation of the signal and the noise as

$$\boldsymbol{R}^l = \boldsymbol{E}_s\boldsymbol{\Lambda}_s E_s^\dagger + \boldsymbol{E}_n\boldsymbol{\Lambda}_n E_n^\dagger,$$

where $\boldsymbol{E}_s = (\boldsymbol{u}(\hat{\xi}_1)^T \boldsymbol{u}(\hat{\xi}_2)^T \cdots \boldsymbol{u}\left(\hat{\xi}_Q\right)^T\Big)$ is a $M \times Q$ matrix,

$$\boldsymbol{\Lambda}_s = \begin{pmatrix} \lambda_1^2 & 0 & \cdots & \cdots & 0 \\ 0 & \ddots & \cdots & \cdots & \vdots \\ \vdots & \vdots & \lambda_p^2 & \cdots & \vdots \\ \vdots & \vdots & \vdots & \ddots & \vdots \\ 0 & \cdots & \cdots & \cdots & \lambda_Q^2 \end{pmatrix}.$$

The control vector from the i th subevent (M subevents in total) can be expressed as

$$\boldsymbol{a}\left(\hat{\xi}_i\right) = [e^{i\omega(t_{i1}-t_{i1})}, \ldots, e^{i\omega(t_{iM}-t_{iM})}]^T.$$

To resolve the arrival direction of the impinging signal, a pseudo-spectrum can be derived with signal directional vectors and noise vectors (Meng et al. 2011) as

$$P_m^l(\xi_i) = \left|\frac{\boldsymbol{a}^\dagger(\xi_i)\boldsymbol{a}(\xi_i)}{\boldsymbol{a}^\dagger(\xi_i)\boldsymbol{E}_n\boldsymbol{E}_n^\dagger\boldsymbol{a}(\xi_i)}\right|.$$

The pseudo-spectrum reaches the maximum when the $\boldsymbol{a}(\xi_i)$ is the directional vector of the real signal. After the azimuth of the signal is determined, the power spectrum can be derived through the signal subspace of the covariance matrix (Mestre et al. 2007) as

$$P_m^l(\xi_i) = \left|\frac{\boldsymbol{a}^\dagger(\xi_i)\boldsymbol{a}(\xi_i)}{\boldsymbol{a}^\dagger(\xi_i)\boldsymbol{E}_s(\boldsymbol{\Lambda}_s - \sigma^2\boldsymbol{I})^{-1}\boldsymbol{E}_s^\dagger\boldsymbol{a}(\xi_i)}\right|. \quad (2.25) \qquad (2.25)$$

For each time step, the location and power of the subevent relative to the hypocenter can be resolved as described above. Thus, the rupture process of the earthquake evolving with time is imaged.

2.2.4 Compressive Sensing

Compressive sensing is a recent technique processing images on the basis of resampling or sampling of data (Donoho 2006; Candes et al. 2006). The principle of the technique is to use data points as few as possible to recover the original signal; this is realized with $\boldsymbol{l}_1$ norm having a special property of sparsity. Before the compressive sensing, $\boldsymbol{l}_p(\boldsymbol{p} \leq 1)$ norm regularization was already widely used to resolve the arrival directions of sparse signals using antenna arrays (Gorodnitsky and Rao 1997; Fuchs 2001; Malioutov et al. 2005). Only limited subfaults slip at some moment during the rupture of an earthquake. This indicates there exists sparsity in the rupture process of the earthquake, which suffices the requirement of compressive sensing. The 2011 Tohoku earthquake is the first one studied by Yao et al. (2011) using compressive sensing.

The procedure to apply the compressive sensing is described below. First of all, the source region is gridded into $\boldsymbol{N}$ subfaults $(\xi_1, \ldots, \xi_N)$. The waveforms radiated from an earthquake are recorded at M stations $(\boldsymbol{x}_1, \ldots, \boldsymbol{x}_N)$. Next, the waveforms in the first time window are aligned to let the epicenter be imagined as the first subevent. The waveforms can be decomposed into plane waves in the frequency. The velocity records at the kth station can be expressed as

$$v(\boldsymbol{x}_k, \omega) = A(x_k, \xi_i, \omega)s(\xi_i, \omega) + n(x_k, \omega),$$

where $v(\boldsymbol{x}_k, \omega)$ is the Fourier transformation of the velocity records at the kth station in frequency domain, $A(\boldsymbol{x}_k, \xi_i, \omega) = e^{i\omega\Delta t_{ik}}$ is phase offset of the waveforms back-projected to the ith subfault relative to the hypocenter ξ_e, $\Delta t_{ik} = t_{ik} - t_{ek}$ is the difference between the travel times of the kth station to the ith subfault and the hypocenter, $s(\xi_i, \omega)$ is the source time function of the i th subfault ξ_i, and $n(\boldsymbol{x}_k, \omega)$ is background noise.

The Eq. 2.50 can be expressed in a vector way as

$$\boldsymbol{v}(\omega) = \boldsymbol{A}(\omega)\boldsymbol{s}(\omega) + \boldsymbol{n}(\omega),$$

where $\boldsymbol{v}(\omega) = [v(\boldsymbol{x}_1, \omega), \ldots, v(\boldsymbol{x}_M, \omega)]^T$, $\boldsymbol{s}(\omega) = [s(\boldsymbol{x}_1, \omega), \ldots, s(\boldsymbol{x}_N, \omega)]^T$, and

$$\boldsymbol{A}(\omega) = \begin{pmatrix} e^{i\omega\Delta t_{11}} & e^{i\omega\Delta t_{12}} & \cdots & e^{i\omega\Delta t_{1N}} \\ e^{i\omega\Delta t_{21}} & e^{i\omega\Delta t_{22}} & \cdots & e^{i\omega\Delta t_{2N}} \\ \vdots & \vdots & \ddots & \vdots \\ e^{i\omega\Delta t_{M1}} & \cdots & \cdots & e^{i\omega\Delta t_{MN}} \end{pmatrix}.$$

The number of subfaults is usually larger than the number of stations, namely $N > M$. This indicates the equation is under-determined and should have infinite solutions. Since there is sparsity in the earthquake source, l_1 norm of the model can be used as regularization. Moreover, the optimal problem with the regularization of the l_1 norm is convex and always has global optimal solutions (Malioutov et al. 2005). Providing that the solutions are sparse enough and the basis is over-complete, the solutions to the optimal problem are stable, even if there are significant noises contaminating the signals. The optimal problem can be constructed as

$$\min\left(\|\boldsymbol{d}(\omega) - \boldsymbol{A}(\omega)\boldsymbol{s}(\omega)\|_2 + \lambda\|\boldsymbol{s}(\omega)\|_1\right),$$

where $\|\cdot\|_1$ and $\|\cdot\|_2$ represent the l_1 and l_2 norms, respectively, and $\boldsymbol{\lambda}$ is regularization factor. The solution to the aforementioned optimal problem can be expressed as

$$\hat{\boldsymbol{s}}(\omega) = \arg\left[\min\left(\|\boldsymbol{d}(\omega) - \boldsymbol{A}(\omega)\boldsymbol{s}(\omega)\|_2 + \lambda\|\boldsymbol{s}(\omega)\|_1\right)\right].$$

Since it is a second-order convex problem, the optimal problem can be solved using interior point solvers (IPS, Boyd and Vandenberghe 2004). The aligned waveforms can be divided into L segments, and each segment has a length of T_w. For each segment, the location and rupture time of the subevent can be obtained by solving the optimal problem described above. Thus, the rupture process of the earthquakes can be imaged.

2.2.5 Surface-Wave Back-Projection Method

Most of the back-projection studies are conducted using high-frequency body waves. For some large earthquakes or local earthquakes, long period surface waves can also be used to image the rupture processes (Yue et al. 2012; Roten et al. 2012). For local or regional earthquakes, the head waves may arrive earlier than the direct waves, and the direct *S* waves are followed by the large-amplitude Lg waves. Using the typical direct *P* waves, the imaging results would be significantly influenced by the other phases. However, surface waves have much larger amplitudes than the body waves. The imaging from the surface waves would not be affected by the preceding waves, though the resolution of the surface-wave imaging is very low.

Surface waves are dispersed, and the surface waves with different frequencies have distinct phase velocities. Wavelet transformation can be applied to extract the surface wave at a certain frequency f_s. The wavelet transformation of the waveforms at the k th station can be expressed as

$$wt(\boldsymbol{x}_k, s, \tau) = \frac{1}{\sqrt{s}} \int_{-\infty}^{+\infty} v(\boldsymbol{x}_k, t)\psi^*\left(\frac{t-\tau}{s}\right)\mathrm{d}t,$$

where $\psi^*\left(\frac{t-\tau}{s}\right)$ is the conjugate of the wavelet and s is the scaling factor.

Without loss of generality, frequency B-spline is used as the wavelet (Teolis 1988):

$$\psi_B(\tau) = \sqrt{f_b}\left[\sin c\left(\frac{f_b t}{p}\right)\right]^p e^{i2\pi f_c t},$$

where f_b is band of the frequency, f_c is the central frequency of the wavelet, p is an integer greater than or equal to 2, and $\sin c(x) = \frac{\sin(x)}{x}, x \in R$. The scaling factor can be obtained with the following equation

$$s = \frac{f_c}{f_s}.$$

For the *i*th subfault, the waveforms after wavelet transferred can be back-projected to the source region and the beam is formed as

$$B(\xi_i, s, \tau, f_s) = \left|\sum_{k=1}^{M} w_{ik} wt(\boldsymbol{x}_k, s, \tau + t_{ik}(f_s))\right|^2,$$

where τ is rupture time, w_{ik} is the weight in terms of the ith subfault and kth station, and $t_{ik}(f_s)$ is travel time of the surface wave with a frequency of f_s. For each time step, the location of the subevent can be obtained by solving the optimal problem as follows:

$$\hat{s}(f_s) = \arg[\max(B(\xi_i, s, \tau, f_s), i = 1, \ldots, N)].$$

Thus, the rupture processes of the earthquakes can be imaged.

2.3 Relative Back-Projection Method

The back-projection methods introduced above have their own features. The first two are similar to the traditional back-projection method, and the second two can be used to identify directions of the signals. The traditional back-projection method is simple, fast, and efficient, but the imaging results severely suffer from the artifacts like "swimming." The latter two approaches are more robust, but may consume more computational resources.

To mitigate the "swimming" artifacts, we modified the traditional back-projection to add a reference station. This is equivalent to add a reference window for the stacking (Meng et al. 2011) and facilitate to suppress the "swimming" artifacts caused by the trade-off between the location and rupture times. The effect of mitigating artifacts will be demonstrated with several synthetic tests in the following section. The improved method is called relative back-projection method (RBPM) due to the reference station used in the method. Moreover, the Pth root stacking technique having more power in suppressing noise was incorporated into the RBPM (Zhang and Ge 2010).

The workflow of the RBPM with the Pth root stacking is presented as follows (Fig. 2.2). First of all, the source region is gridded into N potential subevents. The reference station $\boldsymbol{x}_r$ was used to calculate the difference between the travel times of the ith potential subevent ξ_i to the kth station $\boldsymbol{x}_k$ and to the reference station

$$\Delta t_{ik}^{r} = t_{ik}^{c} - t_{ir}^{c},$$

where t_{ik}^{c} and t_{ir}^{c} are the travel times of the potential subevent ξ_i to the kth station $\boldsymbol{x}_k$ and to the reference station $\boldsymbol{x}_r$, respectively. The time calibration of the kth station $\boldsymbol{x}_k$ is used to compensate the derivation of the one-dimensional velocity model from the heterogeneous earth and is expressed as

$$\delta t_{k}^{r} = t_{k}^{0} - t_{r}^{0} - t_{ek}^{c} + t_{er}^{c},$$

where t_{k}^{0} and t_{r}^{0} are the observed arrival times at the kth station $\boldsymbol{x}_k$ and at the reference station $\boldsymbol{x}_r$, respectively, t_{ek}^{c} and t_{er}^{c} are the theoretical travel times of the epicenter to the kth station $\boldsymbol{x}_k$ and to the reference station $\boldsymbol{x}_r$, respectively. The

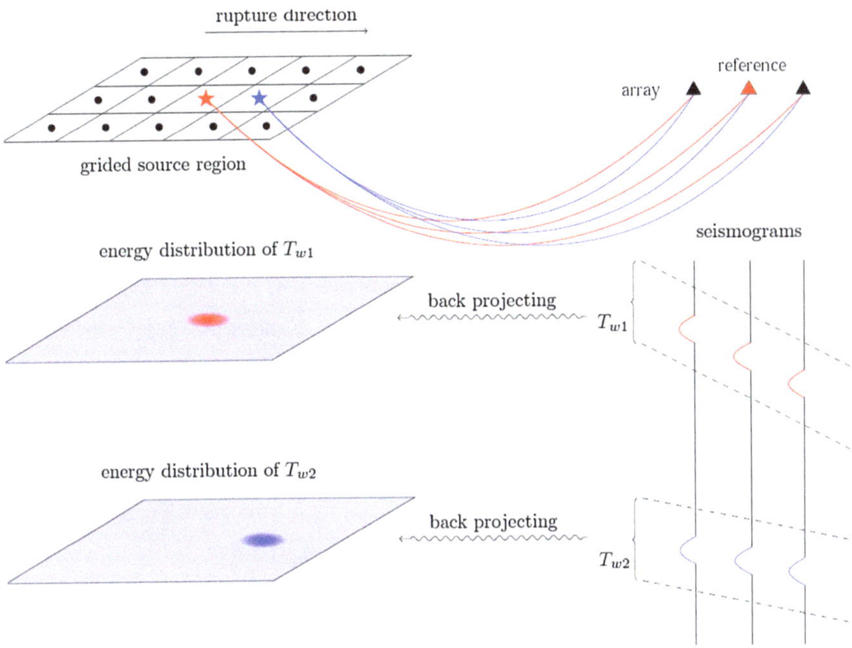

Fig. 2.2 Cartoon showing the back-projection imaging. First, the source region is gridded into smaller blocks (black dots). The red and blue stars indicate two sources at two moments, respectively. The signals generated by the two sources are recorded at the array with a reference station (red triangle). Next, the seismograms are back-projected to the gridded source region with the theoretical travel times. Thus, the power distributions at the two moments can be derived. The powers at the two sources are the maximum (red and blue smearing areas) at the moments when the sources occurred, respectively

velocity records were back-projected and stacked with the P th root stacking to obtain the beam by

$$B(\xi_i, t) = |B'(\xi_i, t)|^{1/N} \cdot \mathrm{sign}\{B'(\xi_i, t)\},$$

where $B'(\xi_i, t) = \frac{1}{M}\sum_{k=1}^{M} |\bar{v}_{ik}^r(t)|^{1/N} \cdot \mathrm{sign}\{\bar{v}_{ik}^r(t)\}$, $\bar{v}_{ik}^r(t) = \alpha_k v(\boldsymbol{x}_k, t + \Delta t_{ik}^r + \delta t_k^r)$, and α_k is the normalization factor of the kth station $\boldsymbol{x}_k$.

Assuming that a large earthquake is composed of L subevents with a duration of T_w, the lth subevent can be imaged based on the subevent having the maximum power. For the lth segment, the power of the beam can be expressed as

$$P_{il} = \frac{1}{T_w} \int_{T_l}^{T_l + T_w} |B(\xi_i, t)|^2 dt,$$

where $T_l = t_r^0 + (l-1)T_s$, and T_s is the time step. The lth subevent makes the power being the maximum

$$P_l^s = \max\{P_{il}, i = 1, \ldots, N\}.$$

The location of the lth subevent can be identified. The rupture time of the subevent can be obtained using the following equation

$$t_k^{rup} = (l-1)T_s + t_{er}^c - t_{lr}^c.$$

Thus, after the locations and rupture times of all the subevents are obtained, the rupture process of the earthquake is imaged.

2.4 Advantages of the Relative Back-Projection Method

Compared with the traditional back-projection method, the relative back-projection method can be used to mitigate the "swimming" artifacts using a reference station. Moreover, the locations and rupture times of the subevents are resolved sequentially. The improvement of the RBPM can be demonstrated by the rupture model in Fig. 2.3. The rupture model is composed of two sources (S1 and S2). The rupture initiates at the S1 and extends to the S2 at 12 s.

The source causing "swimming" artifacts can be clearly demonstrated in left bottom panel of Fig. 2.3. The three potential subevents (S1–S3) in the rupture model are recorded at the same array with different travel time curves. Using the traditional back-projection method, the waveforms are back-projected to generate very distinct beams at the three potential subevents. Since the sources S1 and S2 are real, the beams back-projected at the two sources have the very large amplitudes due to the coherence of the waveforms as shown in Fig. 2.3b. However, when the waveforms are back-projected to the potential source 3, the beam at 12 s has very large amplitude resulted from the S1, which is larger than that of the beam formed at the potential subevent S2. On the basis of the rule of the maximum power to identify the subevent at each time step, the imaged subevent is the S3 instead of the real subevent S2. This clearly demonstrates the "swimming" artifacts caused by the subevents releasing more high-frequency seismic energy and usually move along the ray path of the array to the sources of the earthquake.

The "swimming" artifacts are caused by the larger-energy subevents during the earthquake. To overcome the influence of these subevents to the smaller subevents, adding a reference station can back-project the waveforms from the larger-energy subevents to all the potential subevents around their real rupture times as shown in Fig. 2.3b. Relative to the traditional BPM, the RBPM uses a reference time window to conduct the back-projection. The waveforms composed of signals from S1 and S2 will be back-projected to the potential subevents S1–S3. Due to high similarity of the waveforms from the same subevents, the beam at the potential subevent S1

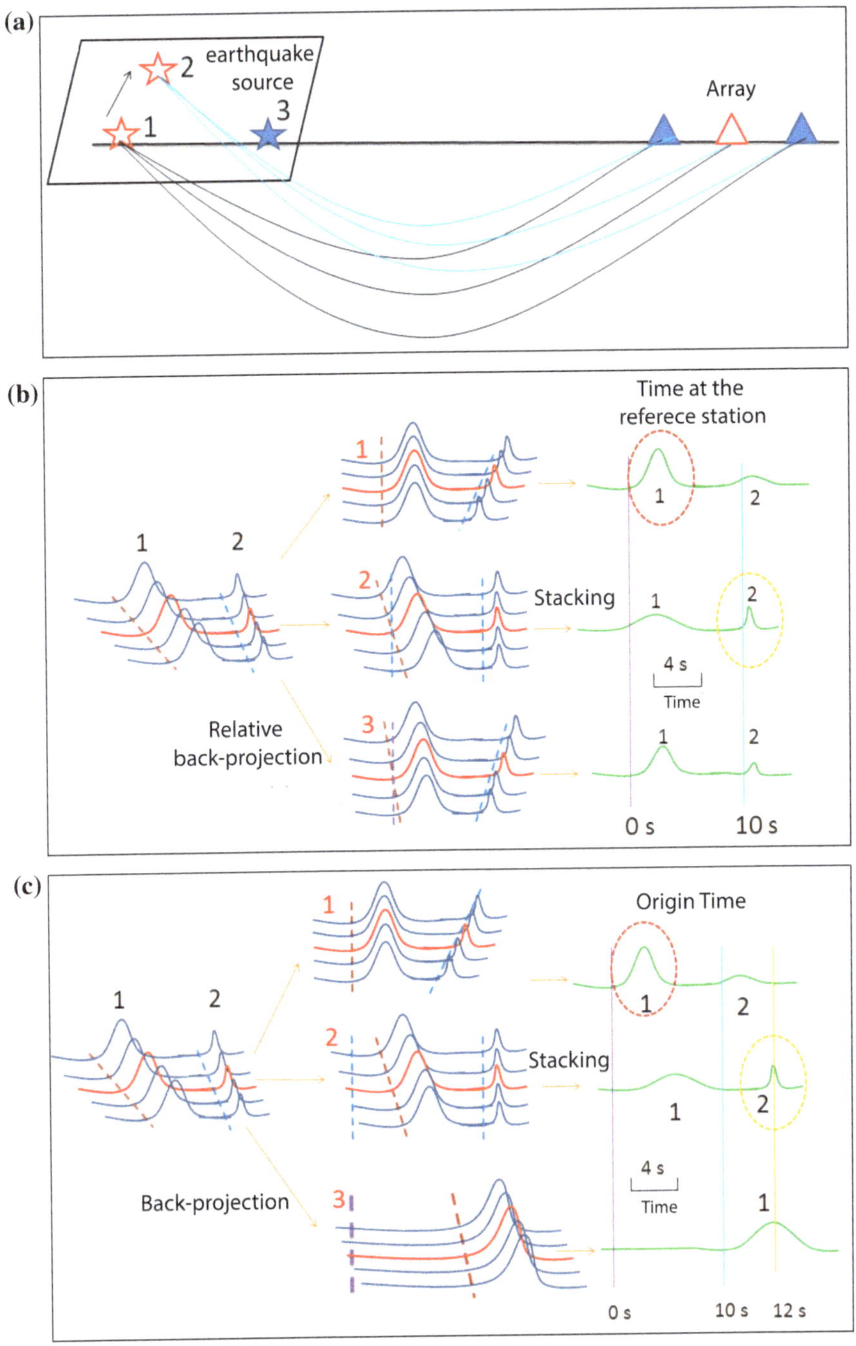

Fig. 2.3 Cartoon showing the difference between the relative back-projection (RBPM) and traditional back-projection (BP) imaging. **a** Two source models. One is composed of sources 1 and 2, and the other one is composed of sources 1 and 3. **b** The imaging procedure of the RBP method. **c** The imaging procedure of the traditional BP method

has the maximum amplitude at 0 s, and the beam at the S2 has the maximum amplitude at 12 s. This means that both S1 and S2 at 0 s and 12 s, respectively can be correctly imaged by the RBPM and evidently demonstrate the effectiveness of the RBPM to suppress the "swimming" artifacts relative to the traditional BPM.

2.5 Synthetic Tests and Comparison

Imaging results by the relative back-projection method (RBPM) have less "swimming" artifacts compared to the traditional back-projection. To verify this point, a series of theoretical tests are conducted as follows. Moreover, the tests directly demonstrate the effective scope of the RBPM and sensitivity of parameters. Specifically, a linear array is used to test the resolutions in the range and azimuth directions. Meanwhile, we also test the influences from reference station, rupture velocity, and filter band. Finally, the rupture imaging results derived by the RBPM are compared to those derived by the traditional BPM.

2.5.1 Unilateral Rupture Tests

The unilateral rupture of an earthquake is simulated with a series of equal-strength explosion point sources for being easy to analyze the results. To obtain the resolutions in the range and azimuth directions, two simulated earthquakes are designed as shown in Fig. 2.4. The first one commences at the eastern end and ruptured westwards. The second one initiates at the southern end and propagates northwards. The receiver array, orientated in a N-S trend, is composed of 201 stations with a spacing of 0.25°, 60° away from the epicenter.

2.5.1.1 The Westward Rupture Earthquake

The earthquake is simulated with eleven explosion sources at a focal depth 0 km, which rupture westwards sequentially and radiate the same seismic moment to each other. The delay time between two neighboring sources 0.05° apart is set according to the rupture velocity of 1 km/s. The specific parameters of these explosion sources as well as the travel times of the direct P waves to the reference station are presented in Table 2.1.

Given the parameters listed in Table 2.1, the seismograms recorded by the linear array can be synthesized using the reflectivity method (Wang 1999) as shown in

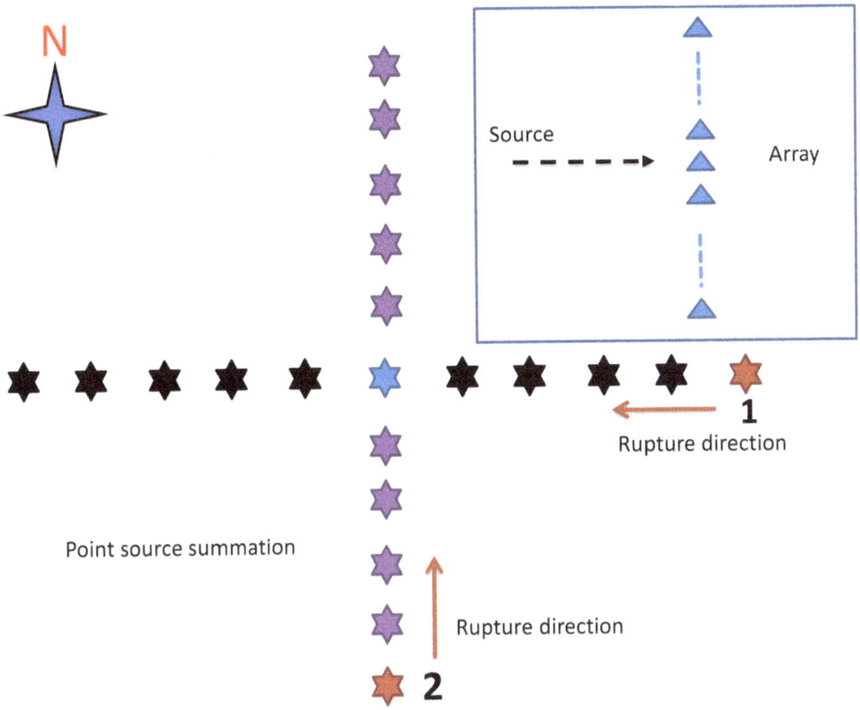

Fig. 2.4 Cartoon of two earthquakes simulated with a series of point sources. The first source model is composed of 11 point sources (black stars) rupturing from east to west, and the second source model is composed of 11 point sources (purple stars) rupturing from south to north. The red stars indicate the epicenters of the two earthquakes. Inset, distribution of a linear N-S striking array, comprising 201 stations (blue triangles)

Fig. 2.5. These vertical-component seismograms are filtered in a frequency band of 0.5–2.5 Hz and aligned with the first *P* arrivals (Fig. 2.5a). The moveouts of the *P* waves generated by the eleven sources are very clear and basically parallel. However, the moveout curve lines from the later ten sources are distorted due to the spatial difference of the ten sources with the first one.

The stacked waveform of the array can be obtained as shown in Fig. 2.5b. It is understandable that the amplitudes of *P* waves from the first source are the largest one close to 1. But the amplitudes of *P* waves from the other sources are all greater than 0.6, though the waveforms are aligned with the *P* waves generated by the first source. This could indicate spatial differences among the point sources may be not very large for the linear array.

The first step of the RBPM is to grid the source region into multiple blocks. The source region centered at the first source is gridded into 101 × 101 blocks with an interval of 0.01° at both the latitude and longitude. The grid points are the potential subevents. The waveforms in a 4-second-long time window are back-projected to these potential subevents with respect to the center station of the linear array every

Table 2.1 The source model for the simulated earthquake, which ruptures westward

Source #	Latitude (°)	Longitude (°)	Origin time (s)	Time (test 1)	Time (test 2)
1	0.00	0.25	0.00	0.00	0.00
2	0.00	0.20	5.57	5.90	5.90
3	0.00	0.15	11.12	11.80	11.70
4	0.00	0.10	16.68	17.70	17.50
5	0.00	0.05	22.24	23.60	23.40
6	0.00	0.00	27.79	29.50	29.30
7	0.00	−0.05	33.35	35.40	35.10
8	0.00	−0.10	38.91	41.30	41.00
9	0.00	−0.15	44.47	47.20	46.80
10	0.00	−0.20	50.03	53.10	52.70
11	0.00	−0.25	55.59	59.00	58.50

Note The parameters of the eleven point sources composing the simulated earthquake 1. The focal depths of all the point sources are set to 0 km. The fifth and sixth columns are the travel times of the sources to the central and southernmost stations, respectively

one second, and the powers are calculated. After the locations of the subevents are determined, their rupture times can be identified accordingly. Thus, the back-projection imaging results are obtained as shown in Fig. 2.6.

Typically, the area outlined by the 80% contour of the maximum power is thought to be the imaging uncertainty (Kruger and Ohrnberger 2005). All the eleven sources are imaged inside their uncertainty areas. The imaging results show that real sources 1–3 are imaged in a diamond area of the maximum power centered at source 1 (Fig. 2.6a, b, c). For source 4, the imaging area with the maximum power extends westward and includes the location of source 4 (Fig. 2.6d). The imaging area of source 4 is composed of two portions. The first one is similar to that of source 4, and the second one is a small area with the location of source 5 (Fig. 2.6e). The imaging area of source 6 mainly focuses on the given location of source 6 (Fig. 2.6f); this indicates the less interference of the waveforms generated by other sources. For sources 7 and 8, they are imaged inside of the diamond areas with the maximum power as well (Fig. 2.6g, h), but the imaging results are affected relatively more by the waveforms from other sources. The maximum imaging area of both sources 9 and 10 includes the real locations of the two sources, but the imaged source 11 is located at edge of the maximum imaging area (Fig. 2.6i, j, k).

These imaging results provide an estimate on the resolution of 22–28 km in the range direction for the linear array 60° away from the epicenter.

The imaging bias in Fig. 2.6 could be caused by the interference from other sources except for source 1. This is the reason why source 1 is imaged correctly. But the multiples, coda waves, or even direct *P* waves generated by the preceding sources could affect the imaging of the latter sources.

The reference station is necessary for the RMPM. In the preceding test, the center station of the linear array is chosen as the reference station. Now, the southernmost station is chosen as the reference station. Again, the same

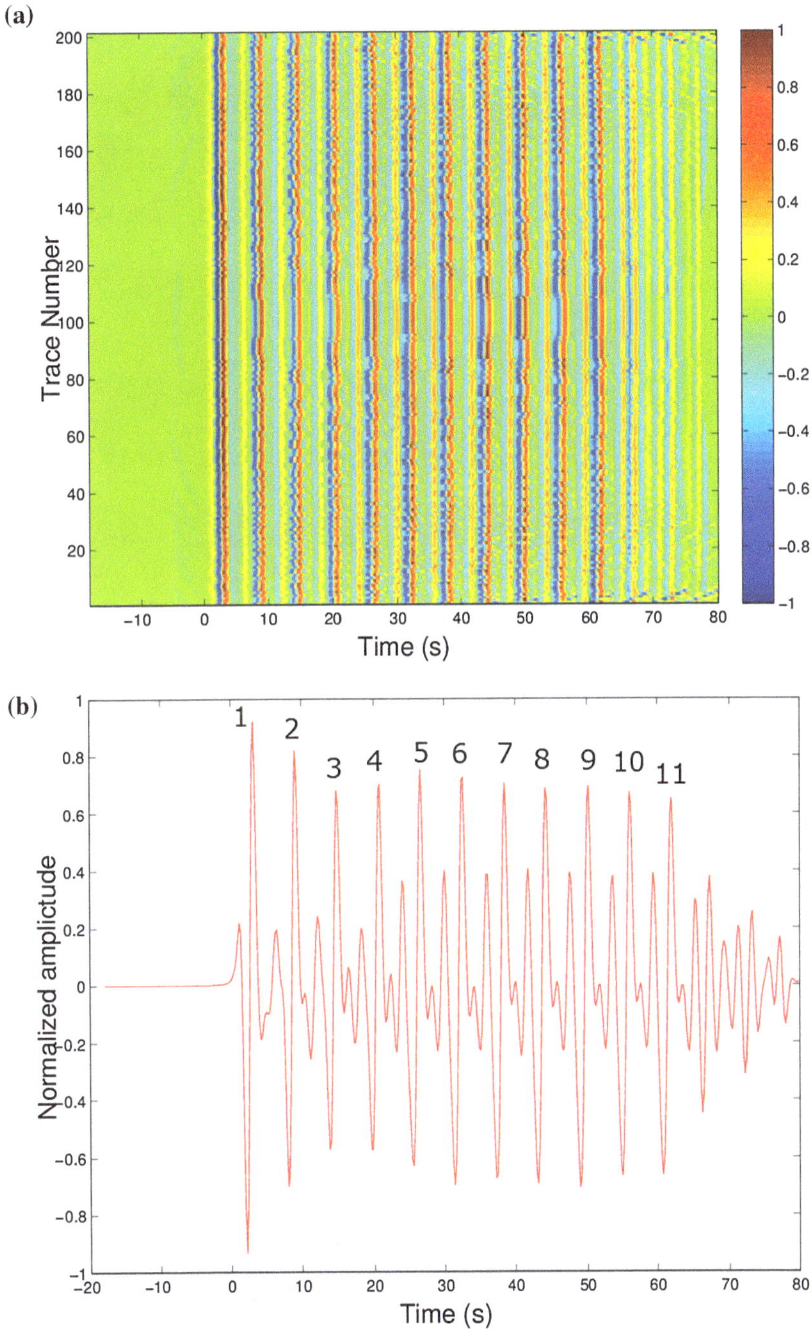

Fig. 2.5 **a** Vertical-component waveforms synthesized for the simulated earthquake 1 and filtered in a frequency range of 0.5–2.5 Hz. The warm color indicates positive amplitudes, whereas the cool color indicates negative amplitude. **b** The beam formed by stacking the waveforms aligned with the first *P* arrivals. The numbers represent the signals from the eleven sources composing the simulated earthquake 1

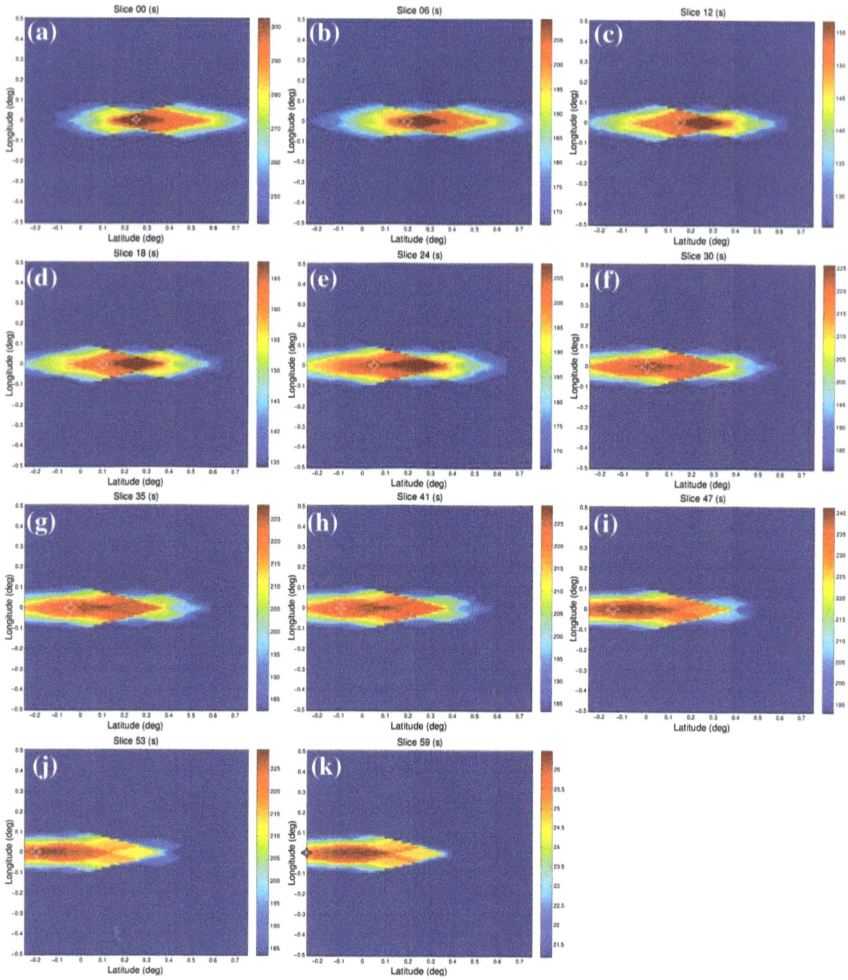

Fig. 2.6 Power distributions derived by the relative back-projection method at the rupture times of the eleven sources composing the simulated earthquake 1. **a–k** Power distribution of the sources 1–11. The star indicates the source. The power is indicated by the color bar with 80% as the color changing indicator

back-projection procedure to the preceding test is conducted, and the imaging results are obtained as shown in Fig. 2.7. Thus, we can do the comparison between the two tests with different reference stations.

In the second test, sources 1–3 are imaged in the maximum imaging areas, and sources 4–6 are imaged just outside the maximum imaging areas. For sources 7–10, they are imaged inside the maximum power area. But source 11 is totally

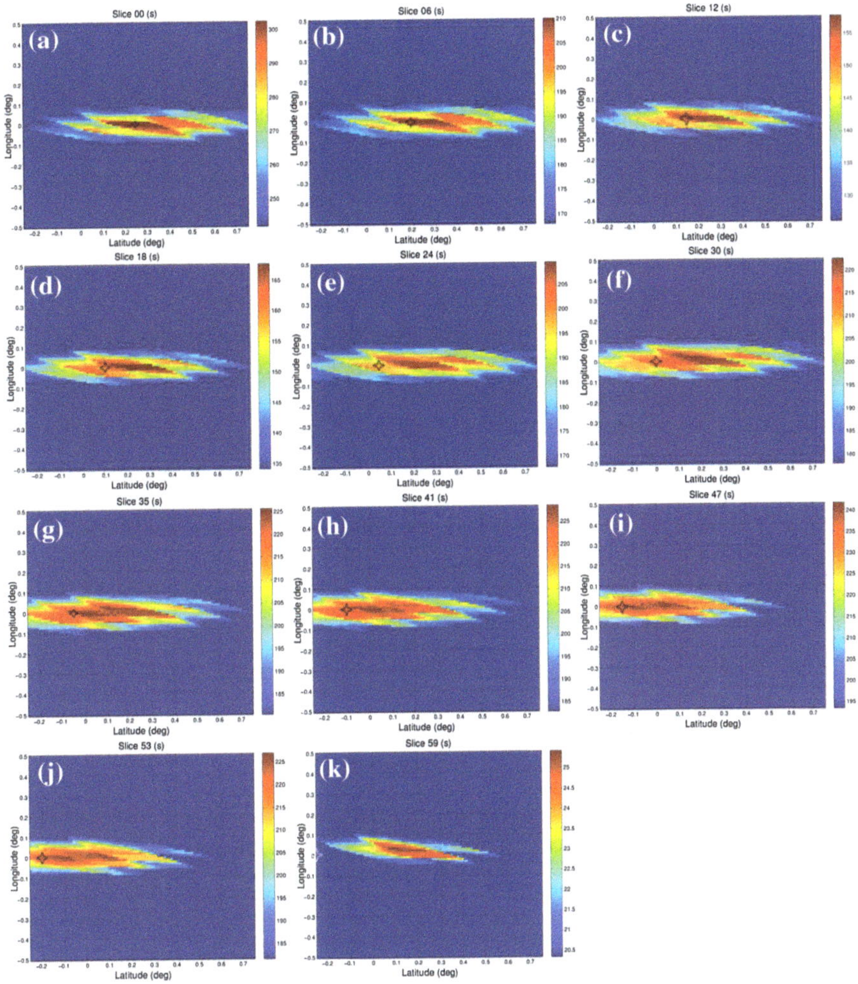

Fig. 2.7 Power distributions derived by the relative back-projection method after changing the southernmost station to the reference one. **a–k** Power distribution of the sources 1–11. The star indicates the source. The power is indicated by the color bar with 80% as the color changing indicator

mis-imaged to an area nearby the locations of sources 1–3. Moreover, the imaging area is elongated in the range direction, along which the spatial resolution is 33 km. This strongly suggests the selection of the reference station does affect the imaging results and the center station is a better choice.

2.5.1.2 The Northward Rupture Earthquake

The northward rupture earthquake is also simulated with eleven explosion source at a depth of 0 km but in a N-S trending orientation. Given a rupture velocity of 1 km/s, the specific parameters of the sources are listed in Table 2.2. With these sources, the vertical-component seismograms are synthesized and filtered in a frequency band of 0.5–2.5 Hz. After they are aligned with the first *P* arrivals of source 1, these waveforms are plotted as shown in Fig. 2.8a. The moveouts of the *P* waves from sources 2–10 are not parallel to that from source 1, which is different from the test on the westward rupture earthquake (Fig. 2.5). This difference is further confirmed by significant decrease in the amplitudes of the stacked waveforms for sources 7–11 (Fig. 2.8b). This indicates the N-S striking array has more spatial sensitive in the azimuth direction than in the range direction.

The *P* waves are back-projected with the same parameters as used in the preceding tests. The imaging results are obtained as shown in Fig. 2.9. The imaging results show that all the sources are imaged at or near the real locations of these sources. The imaging of the later sources is much less affected by the preceding sources relative to the westward rupture earthquake. This agrees with the aforementioned analysis to the synthesized waveforms. The spatial resolution of the N-S linear is about 5.5 km in the azimuth direction, which is much higher than that in the range direction.

2.5.1.3 Frequency-Dependent Test

Since the back-projection imaging has higher resolution for the N-S trending linear array, the northward rupture earthquake is used thereafter to carry out the left tests.

Table 2.2 The source model for the simulated earthquake, which ruptures northward

Source #	Latitude (°)	Longitude (°)	Origin time (s)	Arrival time (s)
1	−0.25	0.00	0.00	0.00
2	−0.20	0.00	5.52	5.50
3	−0.15	0.00	11.04	11.00
4	−0.10	0.00	16.56	16.60
5	−0.05	0.00	22.08	22.10
6	0.00	0.00	27.61	27.60
7	0.05	0.00	33.12	33.10
8	0.10	0.00	38.65	38.70
9	0.15	0.00	44.17	44.20
10	0.20	0.00	49.70	49.70
11	0.25	0.00	55.22	55.20

Note: The parameters of the eleven point sources composing the simulated earthquake 2. The focal depths of all the point sources are set to 0 km. The fifth column is the travel times of the sources to the central station

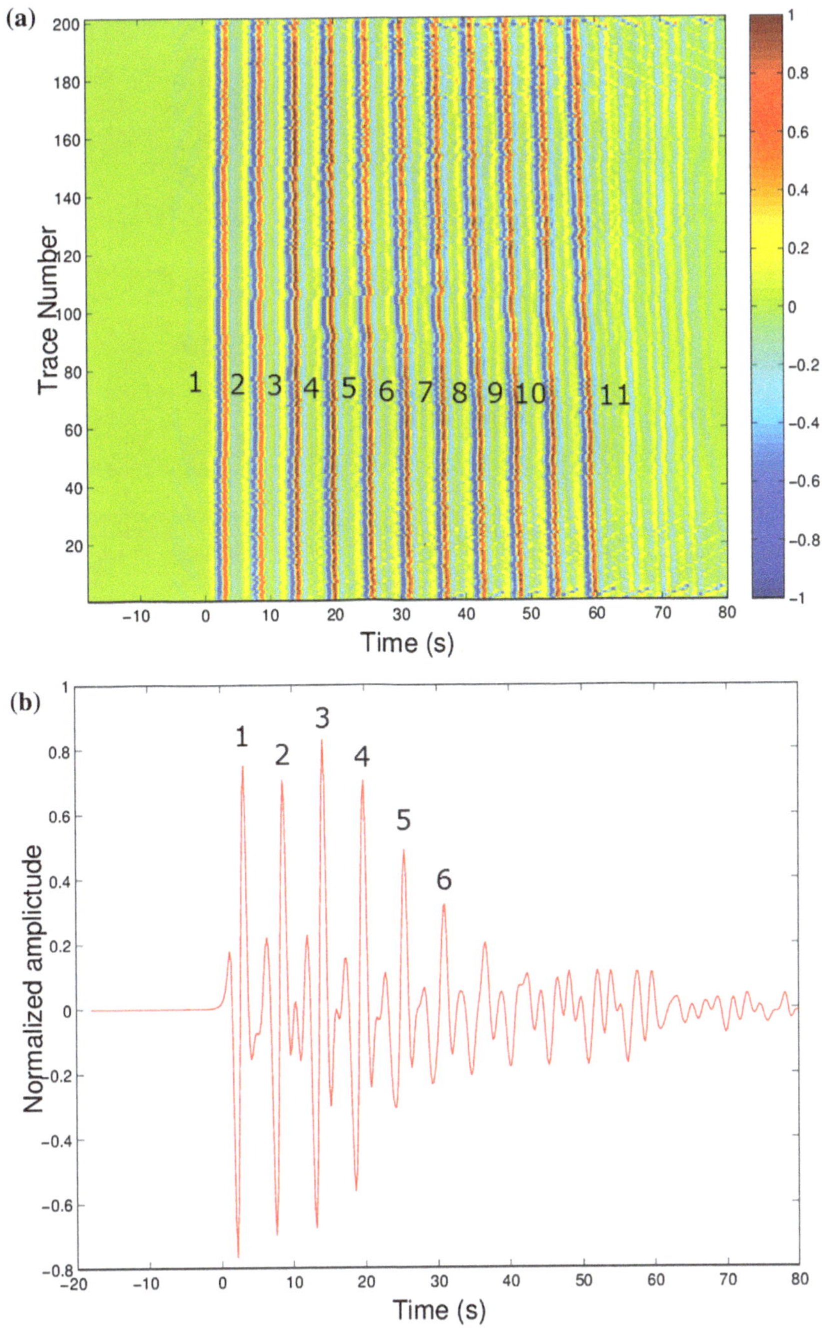

Fig. 2.8 **a** Vertical-component waveforms synthesized for the simulated earthquake 2 and filtered in a frequency range of 0.5–2.5 Hz. The warm color indicates positive amplitudes, whereas the cool color indicates negative amplitude. **b** The beam formed by stacking the waveforms aligned with the first *P* arrivals. The numbers represent the signals from the first six sources composing the simulated earthquake 2

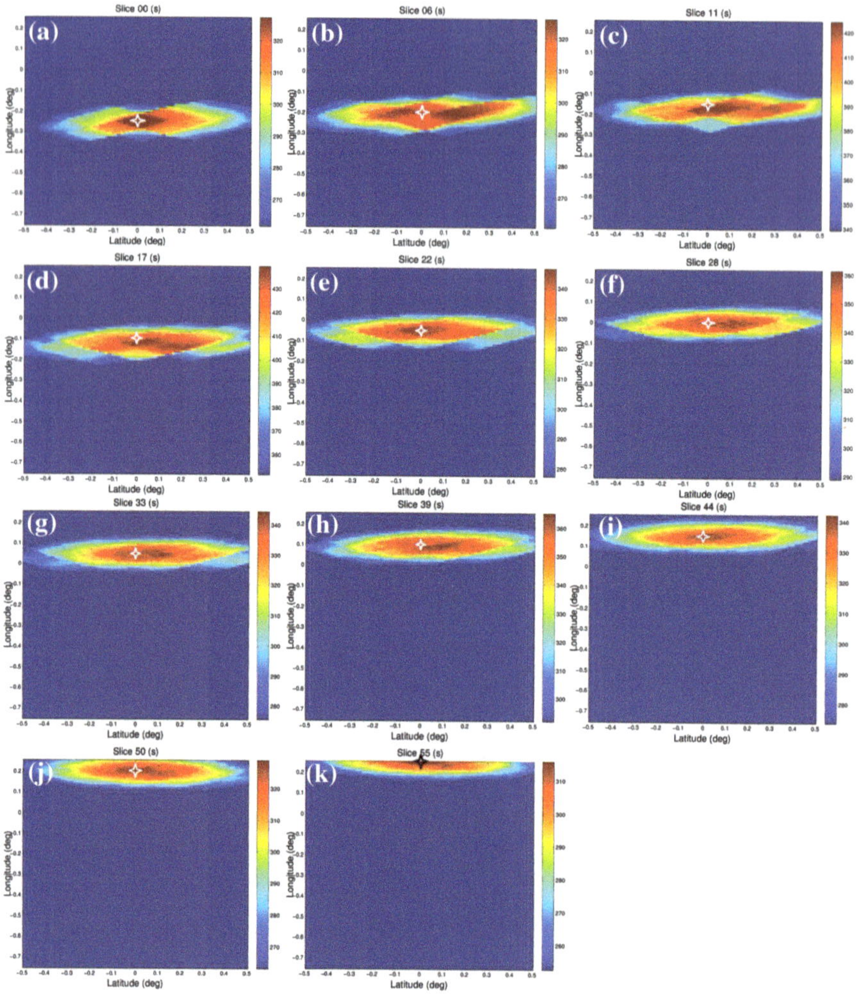

Fig. 2.9 Power distributions derived by the relative back-projection method at the rupture times of the eleven sources composing the simulated earthquake 2. **a**–**k** Power distribution of the sources 1–11. The star indicates the source. The power is indicated by the color bar with 80% as the color changing indicator

Using the back-projection method, a new feature called frequency dependent has been discovered in the rupture of the subduction zone megathrust earthquakes (e.g., Kiser and Ishii 2011; Koper et al. 2011a; Lay et al. 2012; Yao et al. 2013). The frequency-dependent rupture indicates that the high-frequency sources are located deeper than or at the edges of the low-frequency rupture sources. To check whether the imaging results agree with the high-frequency ones, we image the northward rupture earthquake in a low-frequency band of 0.1–0.5 Hz. This facilitates judging whether this frequency-dependency feature is real or caused by the method itself.

The waveforms are simulated with the eleven sources rupturing from south to north at a speed of 1 km/s. The waveforms filtered in the frequency band of 0.1–0.5 Hz are aligned with the first *P* waves as shown in Fig. 2.10a. The longer period of the waveforms makes the interference of the earlier sources to the later sources stronger. Specifically, the amplitudes of the *P* waves generated by the eleven sources decay more gently than the higher-frequency case as shown in the stacked waveform (Figs. 1.8b and 1.10b). As demonstrated in the comparison of the back-projection between range and azimuth directions, this indicates the lower-frequency back-projection has lower resolution.

For the lower-frequency test, the parameters are the same as the higher-frequency back-projection expect for the moving time window with a length of 5 s. These waveforms are back-projected to the source region. The rupture process of the simulated earthquake is imaged as shown in Fig. 2.11.

The imaging results show that all the eleven sources are imaged in the maximum power areas. Specifically, sources 1–4 are imaged at the centers of the maximum power areas (Fig. 2.11a–d), which coincide with the real locations of these sources, respectively. This means these sources are imaged correctly without any bias. However, sources 5–11 are imaged at left edge of the maximum power area. This indicates imaging of these sources is affected by multiples or the direct P waves by the preceding sources. Interestingly, the imaging power is mainly elongated in E-W direction. Thus, the imaging of the sources in the N-S direction is not affected by the interference from other sources. This indicates the frequency band used to filter waveforms would not bias the location the sources if the arrays used have appropriate configurations and enough stations. This strongly supports the argument that the frequency-dependent feature discovered in the rupture of the megathrust subduction zone earthquakes is real (Lay et al. 2012; Yao et al. 2013).

An important thing is that the rise times of the explosion sources are 1 s, which agrees with the frequency band of the data used in the back-projection (0.1–2.5 Hz). For data with frequency beyond this scope, the results of the preceding tests should be applied to do analysis with caution.

2.5.1.4 Test on Rupture Velocity

Rupture velocity is a very important parameter for an earthquake and is typically in a range of 1.0–3.5 km/s. The preceding tests are based on a slow rupture velocity of 1.0 km/s. Next, we will carry out a back-projection test on the northward rupture

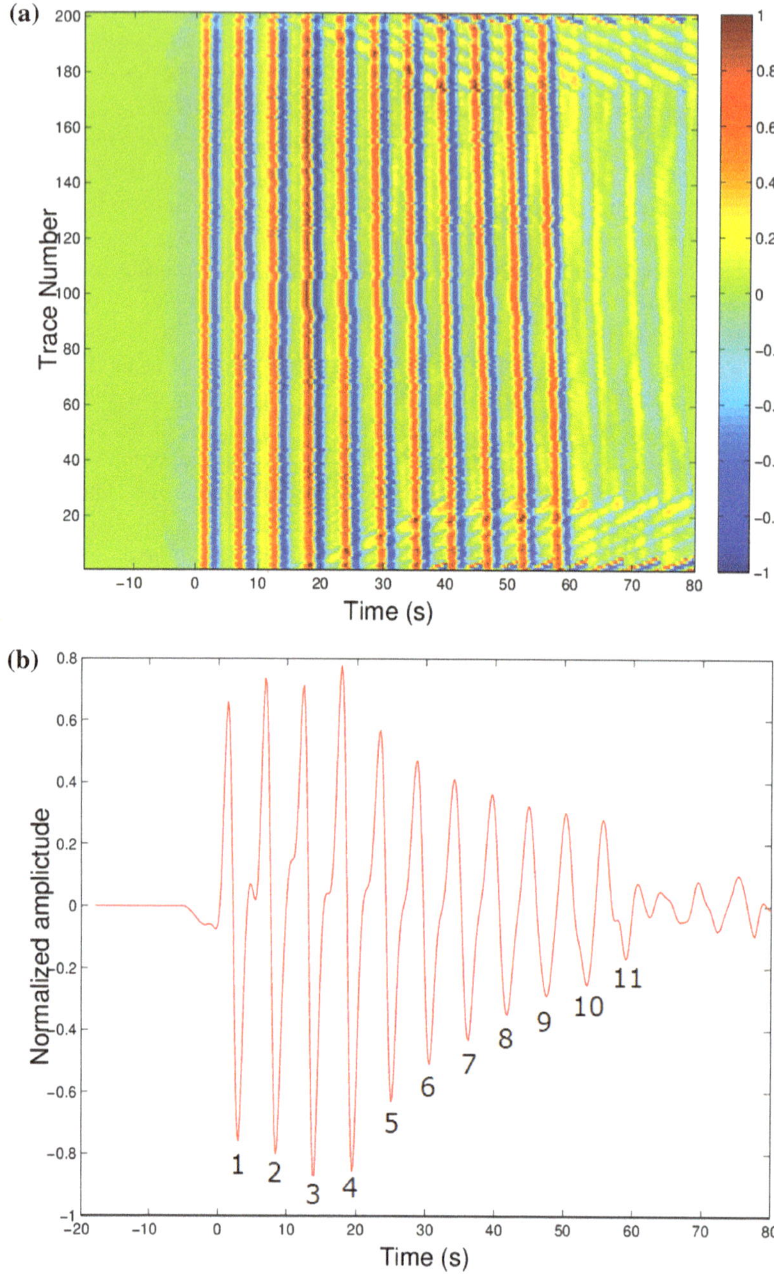

Fig. 2.10 **a** Vertical-component waveforms synthesized for the simulated earthquake 2 and filtered in a low-frequency range of 0.1–0.5 Hz. The warm color indicates positive amplitudes, whereas the cool color indicates negative amplitude. **b** The beam formed by stacking the waveforms aligned with the first *P* arrivals. The numbers represent the signals from the eleven sources composing the simulated earthquake 2

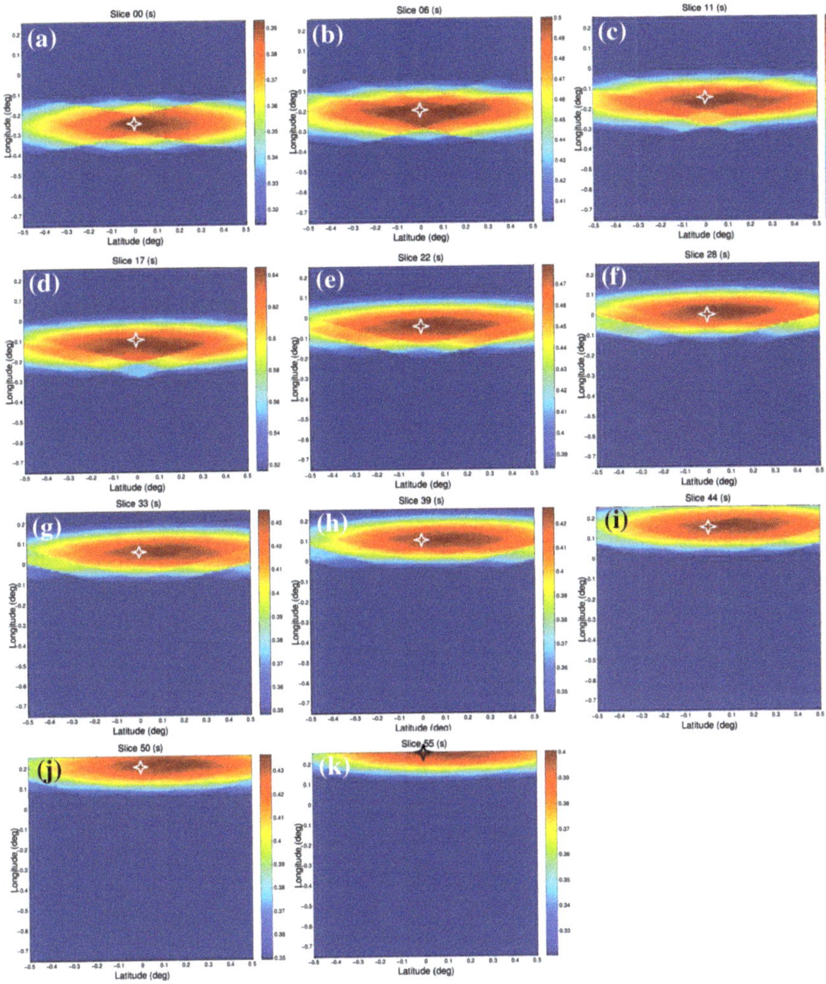

Fig. 2.11 Low-frequency power distributions derived by the relative back-projection method at the rupture times of the eleven sources composing the simulated earthquake 2. **a–k** Power distribution of the sources 1–11. The star indicates the source. The power is indicated by the color bar with 80% as the color changing indicator

earthquake (Fig. 2.4) using a high rupture velocity of 3 km/s. Thus, the comparison between high- and low-velocity cases can test the effect of the rupture velocity on the back-projection imaging.

Using the rupture model comprising the eleven sources, the waveforms are synthesized and then filtered in a frequency band of 0.5–2.5 Hz. After they are aligned with the first time window, the stacked beams are obtained by summing up the aligned waveforms as shown in Fig. 2.12. The *P* arrivals from these sources have less time intervals and the interval becomes shorter and shorter from south to north. Moreover, the multiples have larger amplitudes. Since the time length of the filtered *P* waves is 2–3 s and the rupture time offset between two neighboring sources is ~2 s, the *P* waves from the two sources overlap with each other. The overlapping of the *P* waves could lead to the interference of the imaging.

Since the rupture velocity is faster, the time step is changed to 0.5 s. Other parameters are set to the same. After applying the relative back-projection procedure described above, the rupture of the simulated earthquake can be imagined as shown in Fig. 2.13.

The imaging results show that the *P* waves generated by the early-ruptured sources interfere the imaging of the later ruptured sources. The sources 1–4 are imaged outside of the maximum power areas (Fig. 2.13a–d). This is caused by the interference of the waves from the early-ruptured sources. However, the sources 5–9 except for source 8 are imaged inside their own maximum power areas (Fig. 2.13e–g, i), though the imaging is elongated to E-W direction. Sources 8 and 10 are correctly imaged at the real locations in the N-S direction with a minor westward offset less than 5.5 km (Fig. 2.13h, j). Nevertheless, source 11 is imaged at the location of source 10, which is caused by the interference of the waveforms from source 10.

Although there are larger imaging offsets in E-W direction, the imaging offsets of the sources are less than 5.5 km in the E-N direction. These offsets are resulted from the early-ruptured sources due to the overlapping of waveforms for two neighboring sources. The rupture velocity controls the time offset of waveforms from the two neighboring sources. This indicates the faster the rupture velocity is, the more imaging interferences from the early-ruptured sources are for the later ruptured sources. To avoid these interferences, an appropriate time window needs to be chosen.

2.5.2 Comparison Between the RBPM and Traditional BPM

To carry out the comparison, we exploit the traditional BPM to image the two simulated earthquakes as imaged by the RBPM. The data are filtered in the frequency band of 0.5–2.5 Hz, and the rupture velocity is set to 1.0 km/s.

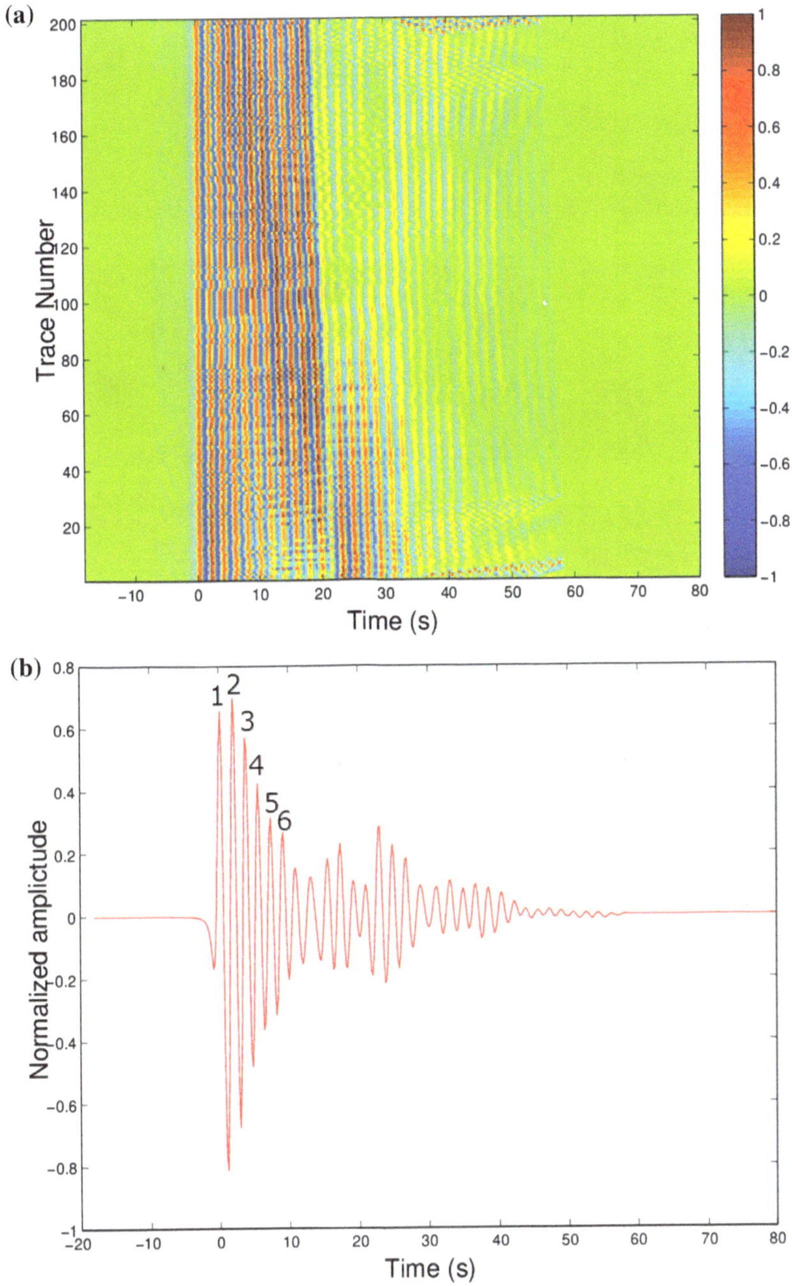

Fig. 2.12 **a** Vertical-component waveforms synthesized for the simulated earthquake 2 and filtered in a frequency range of 0.5–2.5 Hz. The rupture velocity of 3 km/s was used in the simulation of the earthquake. The warm color indicates positive amplitudes, whereas the cool color indicates negative amplitude. **b** The beam formed by stacking the waveforms aligned with the first *P* arrivals. The numbers represent the signals from the first six sources composing the simulated earthquake 2

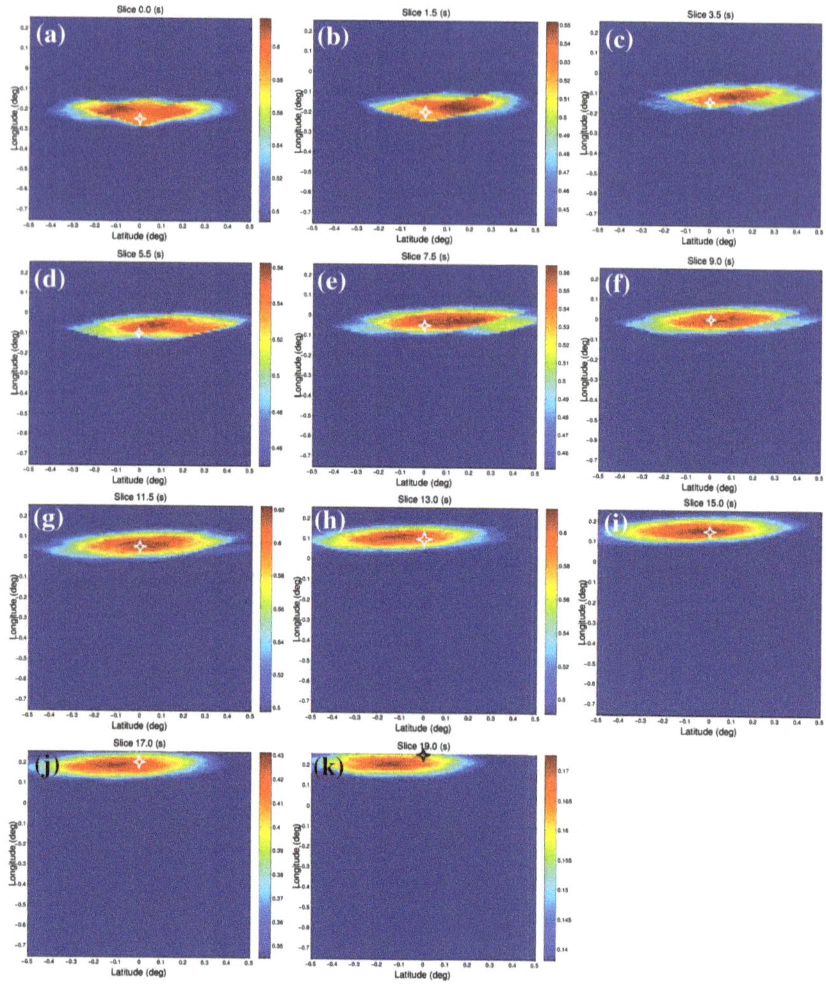

Fig. 2.13 Power distributions derived by the relative back-projection method at the rupture times of the eleven sources composing the earthquake 2 simulated at the rupture velocity of 3 km/s. **a–k** Power distribution of the sources 1–11. The star indicates the source. The power is indicated by the color bar with 80% as the color changing indicator

2.5.2.1 Westward Rupture Earthquake Imaged by the Traditional BPM

According to the preceding analysis of the waveforms synthesized from the western rupture earthquake, the N-S orientated linear array has a lower resolution to image the earthquake relative to the following simulated earthquake. These waveforms are used to image the rupture of the earthquake using the traditional BPM with the same parameters to the RBPM. Thus, the earthquake is obtained as shown in Fig. 2.14.

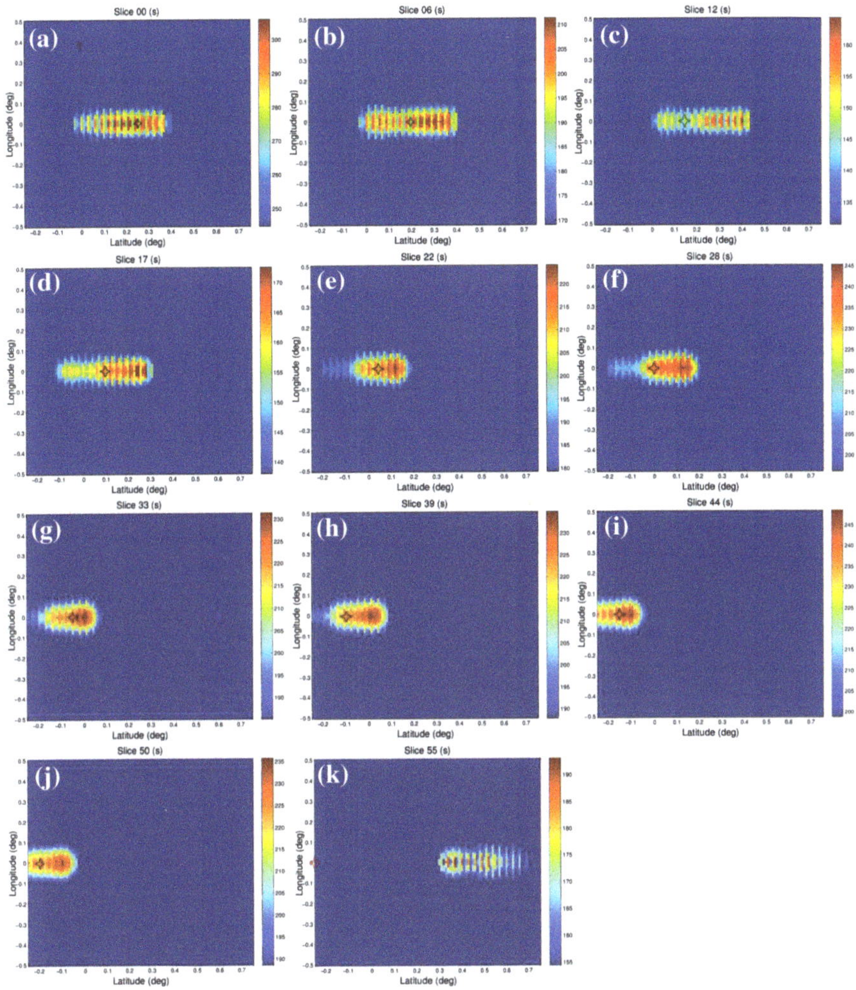

Fig. 2.14 Power distributions derived by the traditional back-projection method at the rupture times of the eleven sources composing the simulated earthquake 1. **a**–**k** Power distribution of the sources 1–11. The star indicates the source. The power is indicated by the color bar with 80% as the color changing indicator

The imaging results show that all the sources are imaged inside 80% contour of the maximum power expect for source 11. Actually, only the imaging maximum power areas of sources 7 and 9 include their real locations, and other sources are outside the maximum power areas. This is the reason why 80% contour of the maximum power is treated as the uncertainty of a source (Kruger and Ohrnberger 2005). Thus, the traditional back-projection has a resolution of 28–44 km along the E-W direction. Moreover, the N-S trending imaging stripes are caused by the array geometry.

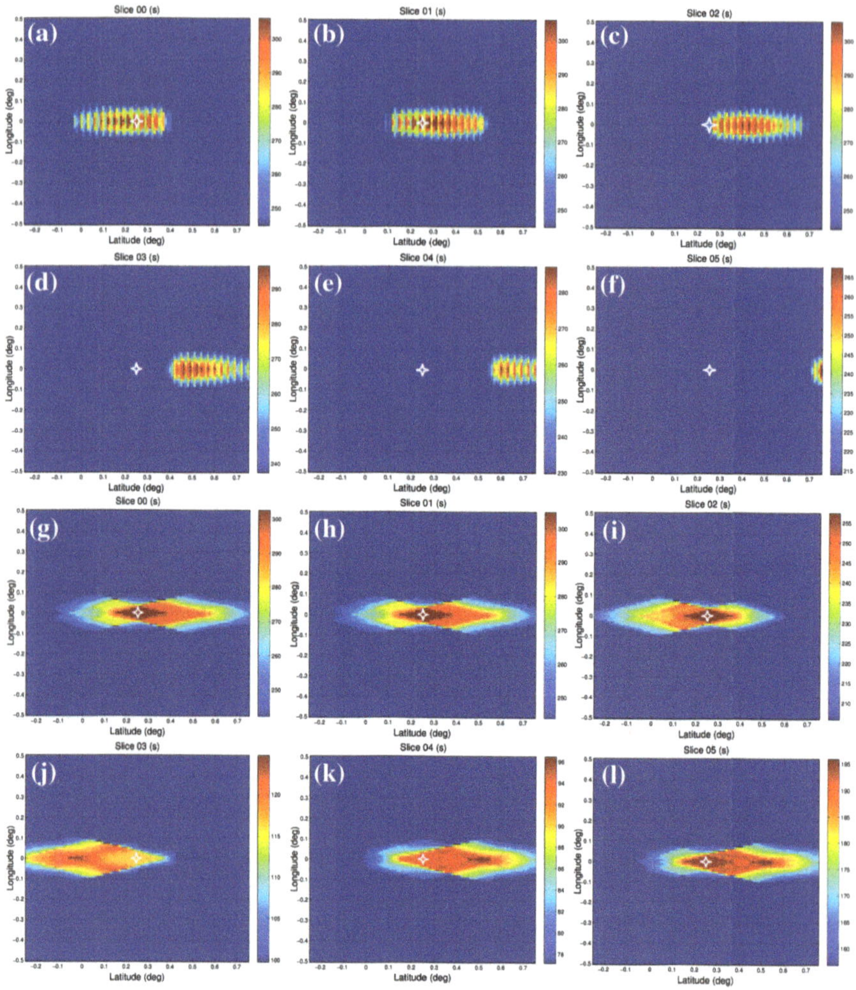

Fig. 2.15 "Swimming" artifacts in the imaging of the simulated earthquake 1 during the first 5 s. **a–f** The traditional back-projection imaging. **g–l** The relative back-projection imaging. The star indicates the source. The power is indicated by the color bar with 80% as the color changing indicator

To avoid missing sources, a time step enough short is used in the imaging. Usually, 1 s is a good choice for the simulated earthquake. However, the "swimming" artifacts clearly show up in the power distribution from 0 s through 5 s as illuminated in Fig. 2.15a–f. In this period, the high-intensity region moves from the center through the east with time. It is just like someone swimming, and this is the reason why this artifact is called "swimming" artifact. The imaging starts again at the center at 6 s (Fig. 2.14b). Thus, the traditional back-projection imaging periodically commences at the real location and then moves toward the array.

For comparison, the relative back-projection imaging is located at the location of source 1 from 0 s through 2 s (Fig. 2.15g–h). However, the maximum power area moves westward at 3–4 s (Fig. 2.15j–k). Two power areas come out at 5 s (Fig. 2.15l): One is located at source 1, and the other one is located the east. Although imaging at 3–4 s deviates from the location of source 1, the maximum powers are less than half of those at 0–2 s (Fig. 2.15g-i). This is because the two neighboring sources are separated ~6 s (Table 2.1) and the time window is 4 s. Starting at 3 s, waveforms in the time window include signals from both sources 1 and 2; this causes the instability of imaging at 3–4 s and lowers down the maximum power due to the incoherency of the waveforms at different stations.

Obviously, the "swimming" artifacts do not exist in the relative back-projection imaging. All the sources are imaged inside 80% contour of the maximum power. As described above, the imaging results from 0 s through 5 s stem from the same source and thus have almost the same maximum powers (Fig. 2.15a–f). This indicates that the "swimming" artifact is a trade-off production of the location and rupture time. Through the comparison, one would know that the RBPM can be used to effectively reduce the "swimming" artifacts.

2.5.2.2 Northward Rupture Earthquake Imaged by the Traditional BPM

Similarly, the second earthquake at a speed of 1 km/s is also simulated (Fig. 2.4), and the waveforms are thus generated as shown in Fig. 2.8a. These waveforms are applied to image the rupture of the earthquake using the traditional BPM. The imaging results (Fig. 2.16) show that although the sources are imaged correctly in the N-S direction, there still exist "swimming" artifacts in the E-W direction.

Overall, after the comparison of the RBPM with the traditional BPM, one knows that the relative back-projection imaging has higher resolution in the range direction and does not have the "swimming" artifacts.

2.5.2.3 Comparison of the Maximum Power

To directly compare the imaging results between the RBPM and traditional BPM, we compile the deviation of the locations with the maximum power from the real locations of the sources for the tests described above. The deviations are obtained for both of the simulated earthquakes as shown in Fig. 2.17. For the first earthquake, all the relative back-projection imaging sources are located closer to the real locations except for sources 7–9 (Fig. 2.17a). For the second earthquake, all the

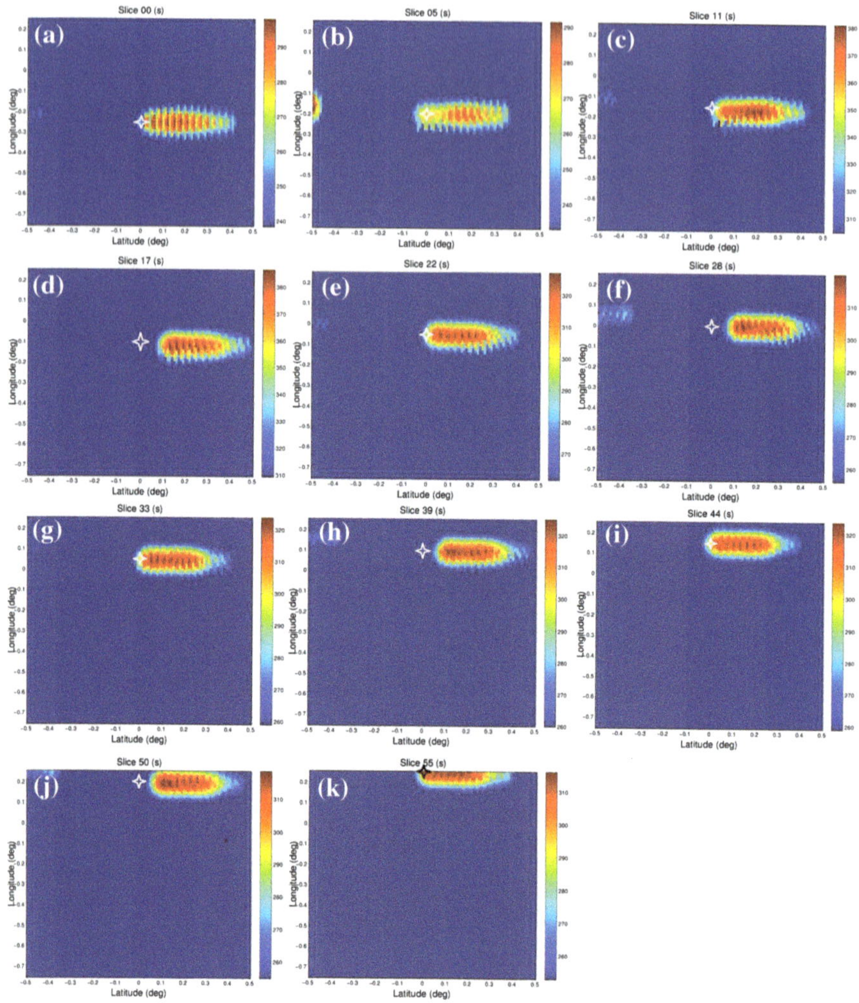

Fig. 2.16 Power distributions derived by the traditional back-projection method at the rupture times of the eleven sources composing the simulated earthquake 2. **a–k** Power distribution of the sources 1–11. The star indicates the source. The power is indicated by the color bar with 80% as the color changing indicator

relative back-projection imaging sources are located closer to the real locations. With the comparison, one can draw a conclusion that the RBPM is a better tool than the traditional BPM to accurately image the rupture of the earthquake with no "swimming" artifacts.

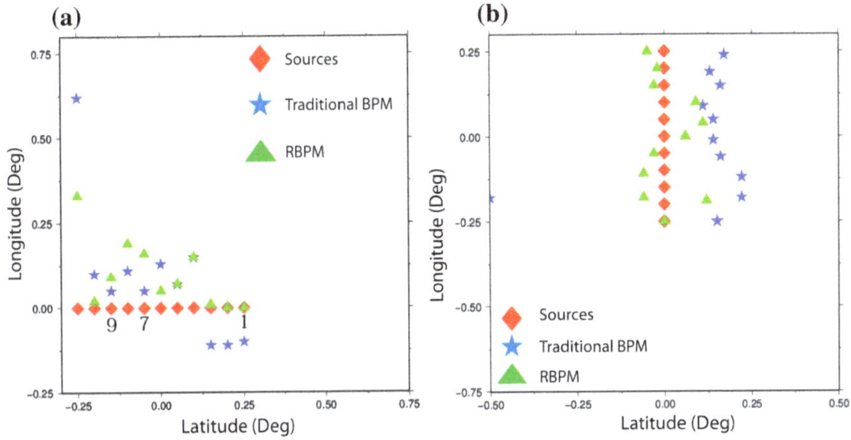

Fig. 2.17 Comparison of the subevents imaged by the relative back-projection and the traditional back-projection methods. **a** The imaging of the simulated earthquake 1. **b** The imaging of the simulated earthquake 2

2.6 Discussion and Summary

In this chapter, the principle of the generalized array techniques was deduced mathematically and it is proofed to be effective and reliable. Five widely used generalized array techniques were then introduced. On basis of the traditional back-projection, an improved approach, namely relative back-projection method (RBPM), was briefly introduced and the workflow to use this method has been elaborated. To verify the reliability of the RBPM, a series of synthetic tests have been conducted. The tests reveal spatial resolutions in range and azimuth directions, and the effects to change important parameters, such as reference station, frequency band used to filter data, and rupture velocity. Finally, a comparison between the RBPM and traditional BPM was carried out to demonstrate the advantage of the RBPM to suppress the "swimming" artifacts in traditional back-projection results.

The principle of the RBPM has been proofed above, though several assumptions have been provided. For example, a large earthquake can be approximated with a sequence of subevents and Green's functions of teleseismic direct *P* waves are larger than multiple's. The first assumption is always sufficed when the event is teleseismic, but the second one is violated when the actual refection waves at surface have larger amplitudes than the direct *P* waves due to radiation pattern. Thus, these assumptions have to be considered first when the RBPM is applied to image the earthquakes.

The RBPM is based on beamforming or slant stacking. When some assumptions are violated, there would be the artifacts in the imaging results. For example, there are "swimming" artifacts in the traditional back-projection imaging results. As demonstrated above, the RBPM can mitigate this kind of artifacts to a great extent. Moreover, the good choice of the reference time and arrays can further improve the imaging.

Table 2.3 The source model for the simulated earthquake, which ruptures northward at a high rupture velocity of 3 km/s

Source #	Latitude (°)	Longitude (°)	Origin time (s)	Arrival time (s)
1	−0.25	0.00	0.00	0.00
2	−0.20	0.00	1.84	1.84
3	−0.15	0.00	3.68	3.68
4	−0.10	0.00	5.52	5.52
5	−0.05	0.00	7.36	7.36
6	0.00	0.00	9.20	9.20
7	0.05	0.00	11.04	11.04
8	0.10	0.00	12.88	12.88
9	0.15	0.00	14.72	14.72
10	0.20	0.00	16.57	16.57
11	0.25	0.00	18.41	18.41

Note The parameters of the eleven point sources composing the earthquake 2 simulated at the rupture speed of 3 km/s. The focal depths of all the point sources are set to 0 km. The fifth column is the travel times of the sources to the central station

The synthetic tests play a key role in understanding the RBPM. Importantly, we test the back-projection imaging using different frequency bands for a given earthquake. The test shows the back-projection results for the two frequency bands basically agree with each other. This confirms the frequency-dependent rupture feature discovered in megathrust subduction zone earthquake (Lay et al. 2012; Yao et al. 2013), indicating the friction varying with depth in the subduction zones. Meanwhile, for linear arrays, the spatial resolution is azimuthal dependent. Moreover, the resolution is also controlled by the rupture velocity and multiples. Overall, the RBPM has the advantage to suppress the "swimming" artifacts relative to the traditional BPM (Table 2.3).

References

Aki, K. 1984. Asperities, barriers, characteristic earthquakes and strong motion prediction. *Journal of Geophysical Research* 89 (B7): 5867–5872.

Aki, K., and P.G. Richards. 2002. Quantitative seismology. University Science Books.

Allmann, B.P., and P.M. Shearer. 2007. A high-frequency secondary event during the 2004 Parkfield earthquake. *Science* 318 (5854): 1279–1283.

Ammon, C.J., C. Ji, H. Thio, et al. 2005. Rupture process of the 2004 Sumatra-Andaman earthquake. *Science* 308 (5725): 1133–1139.

Anstey, N.A. 1964. Correlation techniques–A review. *Geophysical Prospecting* 12 (4): 355–382.

Bolt, B.A., Y.B. Tsai, K. Yeh, and M.K. Hsu. 1982. Earthquake strong motions recorded by a large near-source array of digital seismographs. *Earthquake Engineering and Structural Dynamics* 10 (4): 561–573.

Boyd, S., and L. Vandenberghe. 2004. Convex optimization. Cambridge: Cambridge University Press.

Candes, E., L. Demanet, D. Donoho, and L. Ying. 2006. Fast discrete curvelet transforms. *Multiscale Modeling & Simulation* 5 (3): 861–899.

Chen Y.T., and H.D. Gu. 2008. *Fundementals of earthquake source.*

Das, S., and K. Aki. 1977. Fault plane with barriers: A versatile earthquake model. *Journal of Geophysical Research* 82 (36): 5658–5670.

Donoho, D.L. 2006. Compressed sensing. *IEEE Transactions on Information Theory* 52 (4): 1289–1306.

Du, H.L., L.S. Xu, and Y.T. Chen. 2009. Rupture process of the 2008 great Wenchuan earthquake from the analysis of the Alaska array data. *Chinese Journal Of Geophysics* 52 (2): 372–378. (in Chinese).

Fletcher, J.B., P. Spudich, and L.M. Baker. 2006. Rupture propagation of the 2004 Parkfield, California, earthquake from observations at the UPSAR. *Bulletin of the Seismological Society of America* 96 (4B): 129–142.

Fuchs, J.J. 2001. On the application of the global matched filter to DOA estimation with uniform circular arrays. *IEEE Transactions on Signal Processing* 49 (4): 702–709.

Ge, Z.X., and X.F. Chen. 2008. Point source stacking method to compute synthetic seismogram of finite moving planar source. *Acta Scientiarum Naturalium Universitatis Pekinensis Beijing University (Natural Science)* 44: 407–412.

Goldstein, P., and R.J. Archuleta. 1987. Array analysis of seismic signals. *Geophysical Research Letters* 14 (1): 13–16.

Goldstein, P., and R.J. Archuleta. 1991. Deterministic frequency-wavenumber methods and direct measurements of rupture propagation during earthquakes using a dense array: Theory and methods. *Journal of Geophysical Research* 96 (B4): 6173–6185.

Gorodnitsky, I.F., and B.D. Rao. 1997. Sparse signal reconstruction from limited data using FOCUSS: A re-weighted minimum norm algorithm. *IEEE Transactions on Signal Processing* 45 (3): 600–616.

Hartzell, S.H. 1978. Earthquake aftershocks as Green's functions. *Geophysical Research Letters* 5 (1): 1–4.

Hartzell, S.H., and T.H. Heaton. 1983. Inversion of strong ground motion and teleseismic waveform data for the fault rupture history of the 1979 Imperial Valley, California, earthquake. *Bulletin of the Seismological Society of America* 73 (6A): 1553–1583.

Huang, B.S. 2001. Evidence for azimuthal and temporal variations of the rupture propagation of the 1999 Chi-Chi, Taiwan Earthquake from dense seismic array observations. *Geophysical Research Letters* 28 (17): 3377–3380.

Ishii, M. 2011. High-frequency rupture properties of the Mw 9.0 off the Pacific coast of Tohoku earthquake. *Earth, Planets and Space* 63 (7): 609–614.

Ishii, M., P.M. Shearer, H. Houston, and J.E. Vidale. 2005. Extent, duration and speed of the 2004 Sumatra-Andaman earthquake imaged by the Hi-Net array. *Nature* 435 (7044): 933–936.

Ishii, M., P.M. Shearer, H. Houston, and J.E. Vidale. 2007. Teleseismic P wave imaging of the 26 December 2004 Sumatra-Andaman and 28 March 2005 Sumatra earthquake ruptures using the Hi-net array. *Journal of Geophysical Research: Solid Earth (1978–2012).* 112: B11307.

Kanasewich, E.R., C.D. Hemmings, and T. Alpaslan. 1973. Nth-root stack nonlinear multichannel filter. *Geophysics* 38 (2): 327–338.

Kennett, B.L.N., E.R. Engdahl, and R. Buland. 1995. Constraints on seismic velocities in the Earth from traveltimes. *Geophysical Journal International* 122 (1): 108–124.

Kikuchi, M., and H. Kanamori. 1982. Inversion of complex body waves. *Bulletin of the Seismological Society of America* 72(2): 491–506.

Kiser, E., and M. Ishii. 2011. The 2010 Mw 8.8 Chile earthquake: Triggering on multiple segments and frequency-dependent rupture behavior. *Geophysical Research Letters* 38 (7): L07301. doi: 10.1029/2011GL047140.

Kiser, E., and M. Ishii. 2012. The March 11, 2011 Tohoku-oki earthquake and cascading failure of the plate interface. *Geophysical Research Letters* 39: L00G25.

Koper, K.D., A.R. Hutko, and T. Lay. 2011a. Along-dip variation of teleseismic short-period radiation from the 11 March 2011 Tohoku earthquake (Mw 9.0). *Geophysical Research Letters* 38: L21309.

Koper, K.D., A.R. Hutko, T. Lay, et al. 2011b. Frequency-dependent rupture process of the 2011 M (w) 9. 0 Tohoku Earthquake: Comparison of short-period P wave backprojection images and broadband seismic rupture models. *Earth, Planets and Space* 63 (7): 599–602.

Kruger, F., and M. Ohrnberger. 2005. Tracking the rupture of the Mw & equals; 9.3 Sumatra earthquake over 1,150 km at teleseismic distance. *Nature* 435 (7044): 937–939.

Langston, C.A. 1976. A body wave inversion of the Koyna, India, earthquake of December 10, 1967, and some implications for body wave focal mechanisms. *Journal of Geophysical Research* 81 (14): 2517–2529.

Lay, T., and H. Kanamori. 1981. An asperity model of large earthquake sequences. *Earthquake Prediction* 4: 579–592. (Maurice Ewing Series). doi: 10.1029/ME004p0579.

Lay, T., H. Kanamori, C.J. Ammon, et al. 2005. The great Sumatra-Andaman earthquake of 26 December 2004. *Science* 308 (5725): 1127–1133.

Lay, T., H. Kanamori, C.J. Ammon, et al. 2012. Depth-varying rupture properties of subduction zone megathrust faults. *Journal of Geophysical Research: Solid Earth (1978–2012)* 117: B04311.

Liu, N., Q.F. Chen, F.L. Niu, and Y. Chen. 2007. Imaging the rupture process of the 2004 Sumatra-Andaman earthquake using direct P waves. *Science Bulletin* 52(11): 1312–1316.

Liu, N., F.L., Niu, Q.F. Chen, and Y. Chen. 2010. Imaging the rupture of the 2010 M8.8 Chile earthquake with a broadband seismic array. *Chinese Journal Geophysics* 53(7): 1605–1610. doi: https://doi.org/10.3969/j.issn.0001-5733.2010.07.011.

Madariaga, R. 1977. High-frequency radiation from crack (stress drop) models of earth-quake faulting. *Geophysical Journal of the Royal Astronomical Society* 51 (3): 625–651.

Madariaga, R. 1983. High frequency radiation from dynamic earthquake fault models. *Annales Geophysicae* 1 (1): 17–23.

Maercklin, N., G. Festa, S. Colombelli, and A. Zollo. 2012. Twin ruptures grew to build up the giant 2011 Tohoku, Japan, earthquake. *Scientific Reports*. 2.

Malioutov, D., M. Cetin, and A.S. Willsky. 2005. A sparse signal reconstruction perspective for source localization with sensor arrays. *IEEE Transactions on Signal Processing* 53 (8): 3010–3022.

Meng, L., A. Inbal, and J.P. Ampuero. 2011. A window into the complexity of the dynamic rupture of the 2011 Mw 9 Tohoku-Oki earthquake. *Geophysical Research Letters* 38: L00G07.

Meng, L., J.-P. Ampuero, Y. Luo, et al. 2012a. Mitigating artifacts in back-projection source imaging with implications on frequency-dependent properties of the Tohoku-Oki earthquake. *Earth, Planets, and Space* 64 (12): 1101–1109.

Meng, L., J.-P. Ampuero, A. Sladen, and H. Rendon. 2012b. High-resolution backprojection at regional distance: Application to the Haiti M7. 0 earthquake and comparisons with finite source studies. *Journal of Geophysical Research: Solid Earth (1978–2012)* 117: B04313.

Meng, L., J.-P. Ampuero, J. Stock, et al. 2012c. Earthquake in a maze: compressional rupture branching during the 2012 Mw 8.6 Sumatra earthquake. *Science* 337 (6095): 724–726.

Mestre, X., B.A. Johnson, and Y.I. Abramovich. 2007. Source power estimation for array processing applications under low sample size constraints. In *IEEE International Conference on Acoustics, Speech and Signal Processing IEEE*, Vol. 2, pp. 897.

Muirhead, K.J. 1968. Eliminating false alarms when detecting seismic events automatically. *Nature* 217: 533–534.

Olson, A.H., and R.J. Apsel. 1982. Finite faults and inverse theory with applications to the 1979 Imperial Valley earthquake. *Bulletin of the Seismological Society of America* 72 (6A): 1969–2001.

Roten, D., H. Miyake, and K. Koketsu. 2012. A Rayleigh wave back-projection method applied to the 2011 Tohoku earthquake. *Geophysical Research Letters* 39: L02302.

Rudnicki, J.W., and H. Kanamori. 1981. Effects of fault interaction on moment, stress drop, and strain energy release. *Journal of Geophysical Research: Solid Earth (1978–2012)* 86(B3): 1785–1793.

Satriano, C., E. Kiraly, P. Bernard, and J.-P. Vilotte. 2012. The 2012 Mw 8.6 Sumatra earth-quake: Evidence of westward sequential seismic ruptures associated to the reactivation of a N-S ocean fabric. *Geophysical Research Letters* 39: L15302.

Schmidt, R. 1986. Multiple emitter location and signal parameter estimation. *IEEE Transactions on Antennas and Propagation* 34 (3): 276–280.

Schweitzer, J., J. Fyen, and S. Mykkeltveit. 2002. Seismic arrays in Chap. 4. *New Manual of Seismological Observatory Practice* 1–51. doi: 10.2312/GFZ.NMSOP_r1_ch9.

Shearer, P.M. 1997. Improving local earthquake locations using the L1 norm and wave-form cross correlation: Application to the Whittier Narrows, California, aftershock sequence. *Journal of Geophysical Research* 102 (B4): 8269–8283.

Spudich, P., and E. Cranswick. 1984. Direct observation of rupture propagation during the 1979 Imperial Valley earthquake using a short baseline accelerometer array. *Bulletin of the Seismological Society of America* 74 (6): 2083–2114.

Spudich, P., and L.N. Frazer. 1984. Use of ray theory to calculate high-frequency radiation from earthquake sources having spatially variable rupture velocity and stress drop. *Bulletin of the Seismological Society of America* 74 (6): 2061–2082.

Teolis, A. 1998. *Computational signal processing with wavelets*. Bostan: Birkhauser.

VanDecar, J.C., and R.S. Crosson. 1990. Determination of teleseismic relative phase arrival times using multi-channel cross-correlation and least squares. *Bulletin of the Seismological Society of America* 80 (1): 150–169.

Walker, K.T., M. Ishii, and P.M. Shearer. 2005. Rupture details of the 28 March 2005 Sumatra Mw 8.6 earthquake imaged with teleseismic P waves. *Geophysical Research Letters* 32: L24303.

Wang, R. 1999. A simple orthonormalization method for stable and efficient computation of Green's functions. *Bulletin of the Seismological Society of America* 89 (3): 733–741.

Wang, D., and J. Mori. 2011. Frequency-dependent energy radiation and fault coupling for the 2010 Mw 8. 8 Maule, Chile, and 2011 Mw9. 0 Tohoku, Japan, earthquakes. *Geophysical Research Letters* 38: L22308.

Xu, T., Y. Gu, and F. Yu. 2005. Analysis and Generalization of Convolution Differential and Integral Recursion Formulas. *Journal of Nanjing University of Science and Technology (Natural Science)* 5: 613–615.

Xu, Y., K.D. Koper, O. Sufri, et al. 2009. Rupture imaging of the Mw 7.9 12 May 2008 Wenchuan earthquake from back projection of teleseismic P waves. *Geochemistry, Geophysics, Geosystems* 10(4): 1–17.

Yagi, Y., A. Nakao, and A. Kasahara. 2012. Smooth and rapid slip near the Japan Trench during the 2011 Tohoku-oki earthquake revealed by a hybrid back-projection method. *Earth and Planetary Science Letters* 355: 94–101.

Yao, H., P. Gerstoft, P.M. Shearer, and C. Mecklenbrauker. 2011. Compressive sensing of the Tohoku-Oki Mw 9.0 earthquake: Frequency-dependent rupture modes. *Geophysical Research Letters* 38: L20310.

Yao, H., P.M. Shearer, and P. Gerstoft. 2012. Subevent location and rupture imaging using iterative backprojection for the 2011 Tohoku Mw 9.0 earthquake. *Geophysical Journal International* 190 (2): 1152–1168.

Yao, H., P.M. Shearer, and P. Gerstoft. 2013. Compressive sensing of frequency-dependent seismic radiation from subduction zone megathrust ruptures. *Proceedings of the National Academy of Sciences* 10 (12): 4512–4517.

Yue, H., T. Lay, and K.D. Koper. 2012. En echelon and orthogonal fault ruptures of the 11 April 2012 great intraplate earthquakes. *Nature* 490 (7419): 245–249.

Zhang, H., and Z. Ge. 2010. Tracking the rupture of the 2008 Wenchuan earthquake by using the relative back-projection method. *Bulletin of the Seismological Society of America* 100 (5B): 2551–2560.

Zhang, H., and Z. Ge. 2014. Rupture pattern of the Oct 23, 2011 Van-Merkez, Eastern Turkey earthquake. *Earthquake Science* 27 (3): 257–264.

Zhang, H., and Z. Ge. 2017. Stepover Rupture of the 2014 Mw 7.0 Yutian, Xinjiang, Earthquake. *Bulletin of the Seismological Society of America* 581–591.

Zhang, H., Z. Ge, and L. Ding. 2011. Three sub-events composing the 2011 off the Pacific coast of Tohoku Earthquake (Mw 9.0) inferred from rupture imaging by back-projecting teleseismic P waves. *Earth, Planets and Space* 63 (7): 595–598.

Zhang, H., J. Chen, and Z. Ge. 2012. Multi-fault rupture and successive triggering during the 2012 Mw 8.6 Sumatra offshore earthquake. *Geophysical Research Letters* 39: L22305.

Zhang, H., S. van der Lee, and Z. Ge. 2016a. Multiarray rupture imaging of the devastating 2015 Gorkha, Nepal, earthquake sequence. *Geophysical Reseach Letters* 43: 584–591. https://doi.org/10.1002/2015GL066657.

Zhang, H., S. Lee, E. Wolin, T.A. Bollmann, J. Revenaugh, D.A. Wiens, A.W. Frederiksen, F.A. Darbyshire, G.I. Aleqabi, M.E. Wysession, and S. Stein. 2016b. Distinct crustal structure of the North American Midcontinent Rift from P wave receiver functions. *Journal of Geophysical Research: Solid Earth* 121 (11): 8136–8153.

Zhang, H., K.D. Koper, K. Pankow, and Z. Ge. 2017. Imaging the 2016 MW 7.8 Kaikoura, New Zealand Earthquake with Teleseismic P Waves: A Cascading Rupture Across Multiple Faults. *Geophysical Research Letters* 44. doi: https://doi.org/10.1002/2017GL073461.

Chapter 3
Imaging the 2008 Mw 7.8 Wenchuan Earthquake Using the Relative Back-Projection Method

3.1 Introduction

On May 12, 2008, the Wenchuan earthquake with magnitude Mw 7.9 occurred in the Longmen Shan fault zone, which is located at the eastern margin of the Tibetan Plateau and mainly composed of the Wenchuan–Maoxian fault, the Yinxiu–Beichuan fault, and the Guanxian–Jiangyou fault (Zhang et al. 2008), as shown in Fig. 3.1. Its epicenter is located at (30.986° N, 103.364° E) according to the results obtained by the US Geological Survey (USGS). Because the tectonics of the Longmen Shan fault zone is very complex, the focal mechanisms differed at different portions of the fault. The great earthquake ruptured over a length of about 290 km along the Longmen Shan thrust fault zone to the northeast, starting with mainly reverse faulting with small amounts of right-lateral slip and becoming dominantly right-lateral slip toward the end (Burchfiel et al. 2008). This earthquake caused great property damage and loss of human lives. The rapid and accurate estimation of the rupture process and location of the main energy release points are essential for emergency response.

Because of the large scale and the complexity of the rupture region, the rupture must be described simultaneously from temporal and spatial domains. The most widely used way to determine the temporal and spatial rupture process is by using many kinds of inversion methods based on assumed models (Wang et al. 2008; Zhang et al. 2009). However, these methods need many inversion iterations to obtain reasonable results, which consumes a great deal of processing (CPU) time. To resolve the spatial-temporal rupture process of the 2008 Wenchuan earthquake rapidly, a non-plane-wave assumption-based array technology, namely the relative back-projection method, is implemented in this paper.

Since the beginning of the 1960s, seismic array methods have developed rapidly (Rost and Thomas 2002; Schweitzer et al. 2002). There are lots of methods nowadays, such as the beamforming method (Douglas 1998), the frequency-wavenumber method (Capon 1973; Harjes and Henger 1973), the Vespa process–slant-stacking

H. Zhang, *Imaging the Rupture Processes of Earthquakes Using the Relative Back-Projection Method*, Springer Theses,
https://doi.org/10.1007/978-3-662-55239-1_3

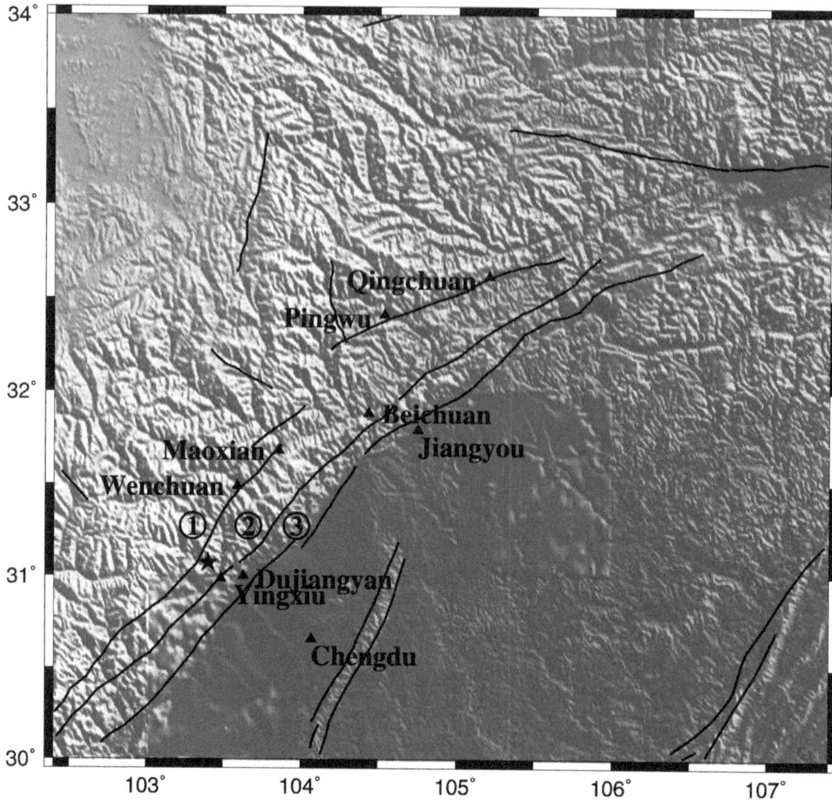

Fig. 3.1 Distribution of the Longmen Shan fault zone and nearby active faults. Location 1 is the Wenchuan–Maoxian fault; location 2 is the Yingxiu–Beichuan fault; and location 3 is the Guanxian–Jiangyou fault. These three faults make up the Longmen Shan fault zone. Asterisk denotes the epicenter of the Wenchuan earthquake

method (Davies et al. 1971), the cophase method (Posmentier and Hermann 1971), the migration method (Lay 1987; Lynnes and Lay 1989), and the back-projection method (Ishii et al. 2005). Most array methods assume the incoming wave is a plane (Birtill and Whiteway 1965). This is a good approximation for teleseismic wave fields (Rost and Thomas 2002) and small-aperture arrays, but it is not suitable for large-aperture seismic network. However, in order to resolve the rupture processes of teleseismic events, a large-aperture seismic network is necessary because the spatial resolution of the array is determined by its aperture (Schweitzer et al. 2002). The plane-wave array technique is of little value for large-aperture seismic networks. Thus, non-plane-wave array methods must be introduced to present the rupture process. The relative back-projection method is one of the methods based on non-plane-wave assumptions. Tracing the rupture process of a great earthquake by using array technology originated from studies of the 2004 Sumatra earthquake (Ishii et al. 2005; Krüger and Ohrnberger 2005). Since then, this kind of method had been used as a regular way to study properties of seismic sources (Walker et al. 2005; Du et al. 2009; Xu et al. 2009).

In this paper, the relative back-projected method based on a regional seismic network is implemented in a two-step scheme. Compared to similar methods used in the previously listed papers, the relative back-projected method presented in this paper has two advantages. First, three-component data instead of vertical-component data are used to estimate the rupture process of the earthquake because the aperture of the seismic networks is large. Second, we used a two-step scheme to determine the time of the rupture fronts, which is necessary to derive the rupture velocity of the earthquake. Due to the finite rupture-propagating speed, the time delay on the seismogram cannot be considered as the time delay of the rupture front, which is not mentioned in those papers. The method is then applied to estimate the two-dimensional, time-dependent rupture properties of the 2008 Wenchuan earthquake in a rapid way, which provides a new perspective for identifying the rupture features of the great Wenchuan earthquake.

3.2 Data and Methodology

The finite source can be simulated by point source stacking (Ge and Chen 2008). The source of a great earthquake can be divided into a series of subevents with locations and rupture times that are related to the rupture fronts. Most of the high-frequency components of the seismic signals were generated from the rupture fronts (Ni et al. 2005). The main goal of the relative back-projection method is to obtain the locations of the maximum high-frequency energy radiation sources as they vary with rupture time.

In order to identify the rupture process of the 2008 Wenchuan earthquake, we use the three-component broadband data of 16 stations in the Australian seismic network. The scheme of applying the relative method to process the data is shown in Fig. 3.2 and described in detail herein.

Distribution of the stations is shown in Fig. 3.3. The epicentral distance of the reference station (STKA in Fig. 3.3) is 72.54°, and the ray path is almost perpendicular to the strike of the fault. Considering the high-frequency assumption for rupture fronts, time resolution, and signal-to-noise ratio, the signals were filtered using a band-pass filter with two corner frequencies (0.4 and 2 Hz). Only the vertical-component signals of the stations are shown in Fig. 3.4. Despite the huge aperture (about 3670 km), the seismograms are very similar to each other. To find the exact locations of the rupture fronts, the direct *P* wave was used. To reduce interference from other seismic phases, mainly *PP*, and include all useful information of the rupture fronts, the direct P wave of the reference station with a length of 160 s was examined.

For the purpose of locating the rupture front, the source region (30.0–35.0° N, 102.4–107.4° E) was spatially gridded into intervals of 0.1° × 0.1° along latitude and longitude, resulting in 2601 potential subevents at a fixed depth of 20 km

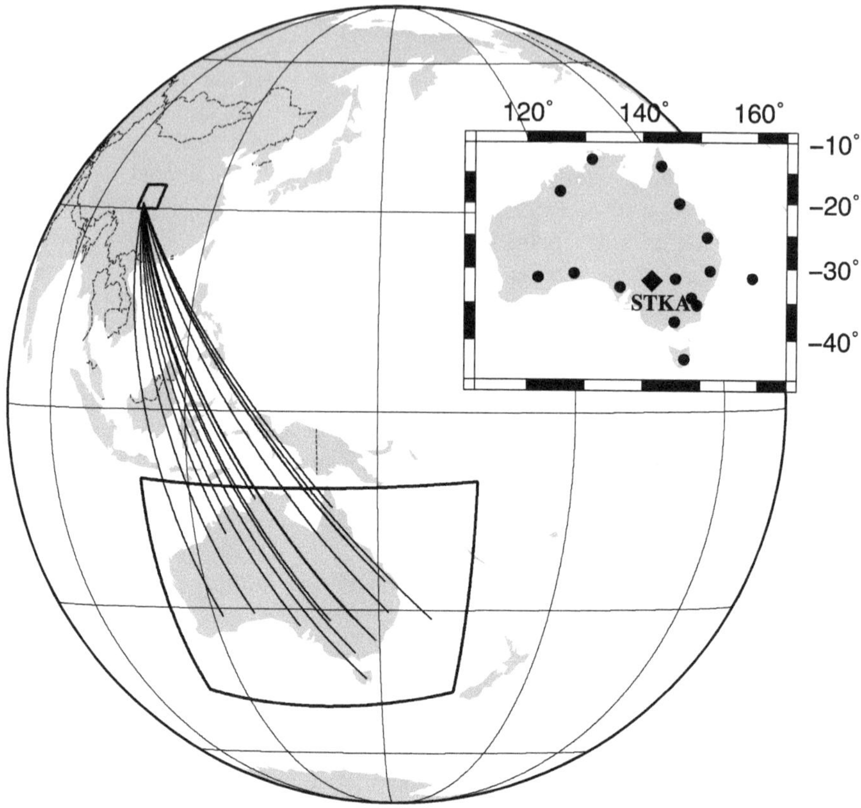

Fig. 3.2 Station distribution of the Australian seismic network. Curved lines, ray paths; circles, stations; and diamond, reference station (STKA)

(which corresponds to the source depth given by USGS). For each grid point, the travel times of direct *P* waves to each station were calculated with the IASPEI91 model. For grid point *i*, the *j*th component signals at *k*th station were normalized and stacked to form a beam:

$$B_j\left(\vec{\xi}_i, t\right) = \frac{1}{M}\sum_{k=1}^{M} \alpha_{kj} v_j\left(\vec{x}_k, t + \delta t_{ik}^c + \Delta t_k\right), \tag{3.1}$$

where $\alpha_{kj} = \frac{p_{kj}}{A_k^{\max}} w_k$, p_{kj} is the polarity of the *P* waves on the *j*th component at the station $\vec{x}_k$, $A_k^{\max}$ is the maximum absolute amplitude of the three-component waveform at the station $\vec{x}_k$, and w_k is the weight of the *k*th station, $v_j(\vec{x}_k, t)$ is the *j*th component velocity waveform at the station $\vec{x}_k$, δt_{ik}^c is the calculated travel time of the *i*th potential subevent to the station $\vec{x}_k$ with respect to the reference station, and Δt_k is the initial travel time correction compensating the travel time derivation of the one-dimensional velocity model from the three-dimensional heterogeneous earth.

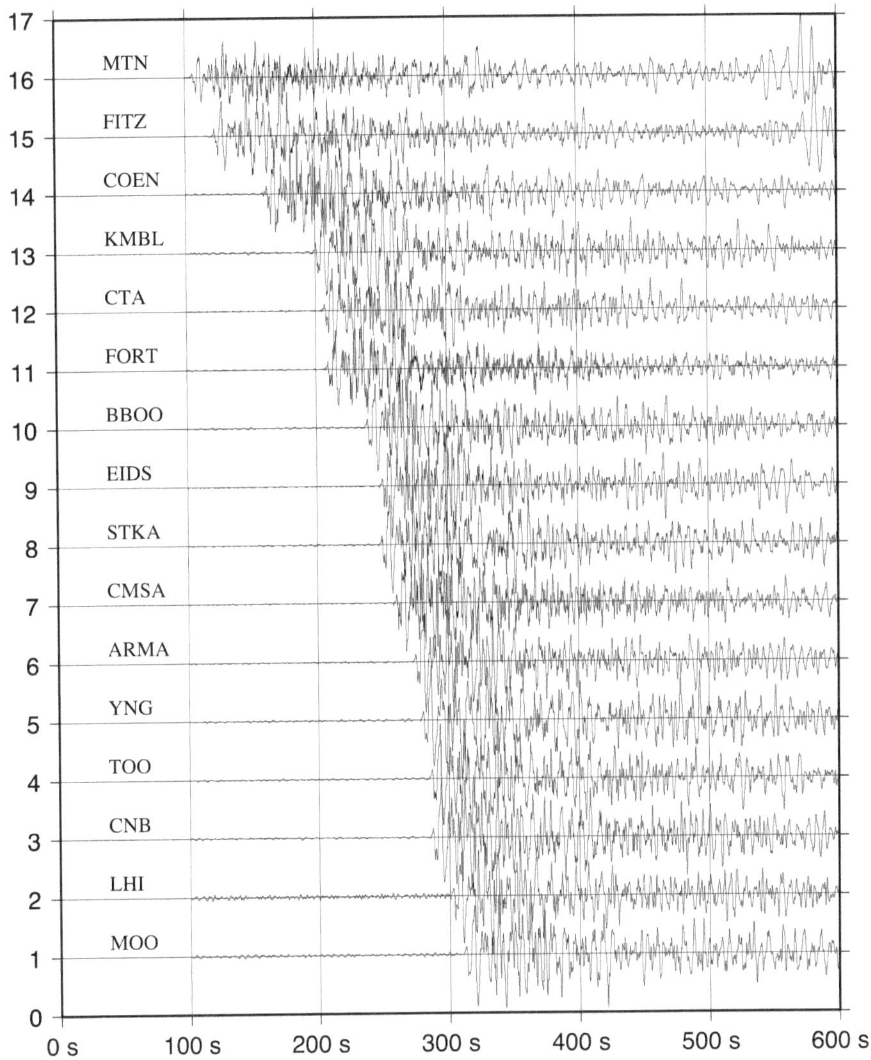

Fig. 3.3 Vertical component of the seismograms recorded by the stations used in this study and sorted by distances to the epicenter

In order to obtain the time correction of the *k*th station, assuming the initial energy is generated at the epicenter, we first calculate the correlating time difference of the first 15 s of seismograms between the *k*th station and the reference station. Then, we calculated the travel time difference between the travel times of the *k*th station to the epicenter and the reference station to the epicenter. Thus, the time calibration of the *k*th station $\vec{x}_k$, compensating the derivation of the one-dimensional velocity model from the heterogeneous earth, is expressed as

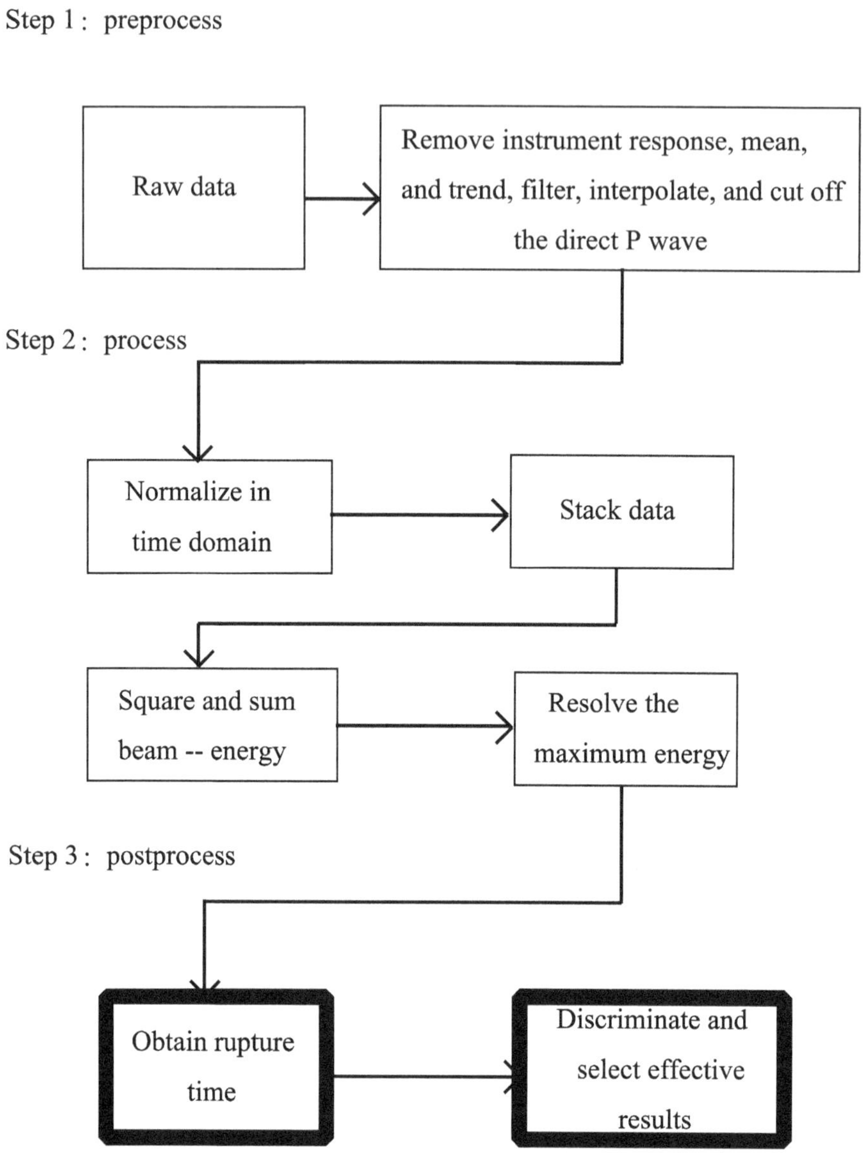

Fig. 3.4 Implementation of the relative back-projection method. Step 1 shows the steps involving in preprocessing data to reduce the influence of instruments and circumstance. Step 2 outlines the processing of data with almost equal magnitude. Step 3 shows the selection of proper results after the discrimination

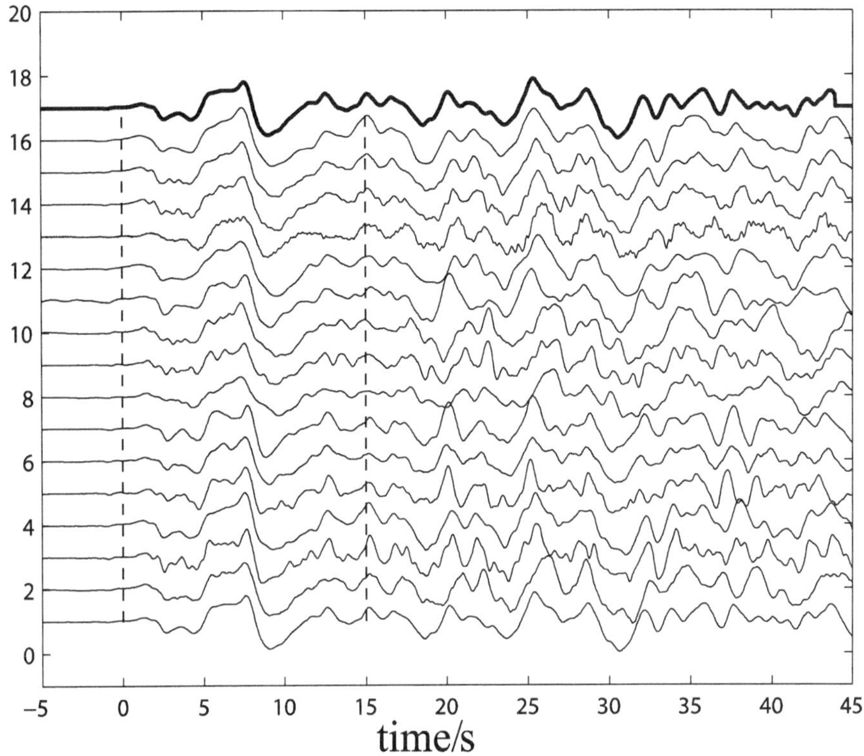

Fig. 3.5 Time-shifted seismogram in 15-s-long segments for all 16 stations. Thick line is the stacked trace

$$\delta t_k^r = t_k^0 - t_r^0 - t_{ek}^c + t_{er}^c, \tag{3.2}$$

where t_k^0 and t_r^0 are the observed arrival times at the kth station $\vec{x}_k$ and at the reference station $\vec{x}_r$, respectively, t_{ek}^c and t_{er}^c are the theoretical travel times of the epicenter to the kth station $\vec{x}_k$ and to the reference station $\vec{x}_k$, respectively. When all the signals were time corrected and synchronized together, the beams of all source grid points could be derived, as shown in Fig. 3.5.

3.2.1 Power of Potential Subevents

The power of potential subevents was calculated in 15-s-long time windows, where the time window started at the first arrival time of the reference seismogram. It moved with a 1-s time interval until the endpoint at 160 s. For the *l*th time window, power of the *i*th potential subevent is defined as

$$P_{il} = \frac{1}{T_w}\sum_{j=1}^{3}\int_{T_l}^{T_l+T_w}\left|B_j(\xi_i,t)\right|^2 dt, \tag{3.3}$$

where $T_l = t_r^0 + (l-1)T_s$, T_s is the time step, and $N = 3$ (the number of components for a station). It is sufficient to use only vertical-component seismograms for a small-aperture array, but three-component seismograms may be needed for a large-aperture array composed of stations in a regional seismic network. The energy map of the grid points of the first time window is shown in Fig. 3.6. As shown, the spatial resolution of the seismic network in the direction along the ray path is lower than that in the direction of the strike of fault. That is the reason why we choose the Australian seismic network.

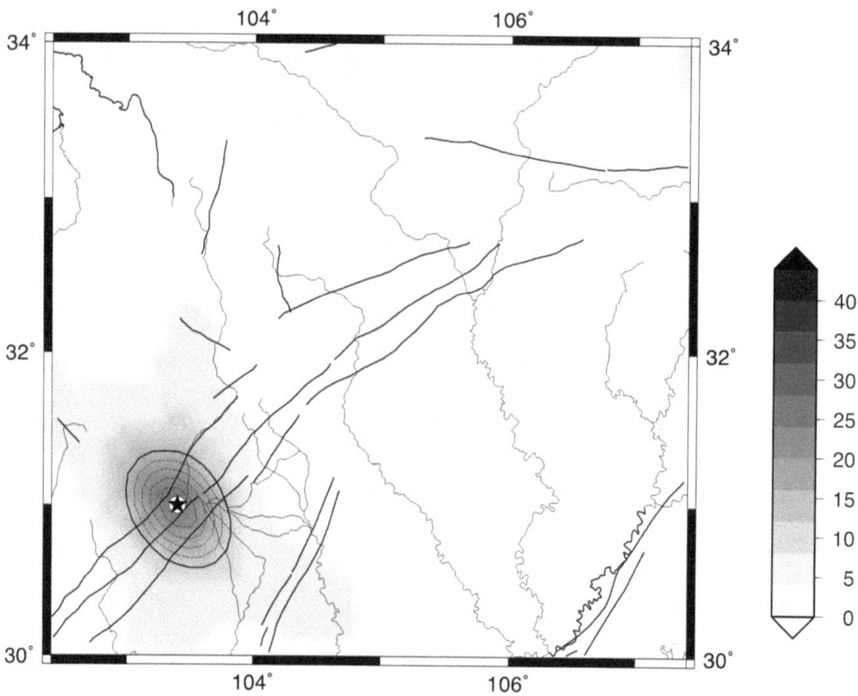

Fig. 3.6 Relative energy distribution of the first time window in the source region. Black lines represent faults, and the star locates the epicenter. Color bar denotes the energy of all grid sources during the entire rupture process after the dataset was normalized. Dashed curved lines indicate the energy contours, with intervals of 10%. The thick solid curved line indicates the contour of 50% of the maximum energy of this time window

3.2.2 Determination of Location and Rupture Time

Two steps were used in determining the location and time of the rupture front. First, the location of the rupture front was determined for each time window. The time window moved 144 steps; for each step, we could get the energy map in the source region. As an example, the energy map corresponding to the initiation of the rupture of the Wenchuan earthquake is shown in Fig. 3.6. We found the maximum energies and the corresponding locations in each time window by

$$P_l^s = \max\{P_{il}, i = 1, \ldots, I\}, \tag{3.4}$$

where I is the number of the potential subevents.

Once the locations were determined, the time corresponding to the subevent could be calculated by subtracting the travel time from the time of the reference station. When location of the epicenter was known and the initial rupture time point was set to be 0 s, rupture time of the nth time window is described by Eq. 3.5 and shown in Fig. 3.7.

$$t_k^{rup} = (l-1)T_s + t_{er}^c - t_{lr}^c, \tag{3.5}$$

where t_{er}^c and t_{lr}^c are the theoretical travel times of the reference station to the epicenter and to the lth subevent station, respectively.

3.2.3 Discrimination of Results

Up to this point, rupture time points and locations of 145 rupture fronts were determined (Fig. 3.8). Because some parts of the signal were very strong, a continuous subset of the rupture fronts with the same location is present in Fig. 3.8. To determine the real rupture time of the signal, only the front with the maximum energy in the continuous rupture fronts was preserved, and the others were removed. Then, we had to discriminate whether the remaining rupture fronts were signal or noise sources. In this paper, a discriminant factor (Du et al. 2009) was used.

$$\gamma = \alpha\beta\kappa, \tag{3.6}$$

where α denotes the correlation coefficient between one time window length signal of the beam and the corresponding reference signal; β denotes amplitude ratio; and κ denotes the energy ratio of the maximum energy of one time window to the maximum energy of the whole duration of 160 s (Fig. 3.9).

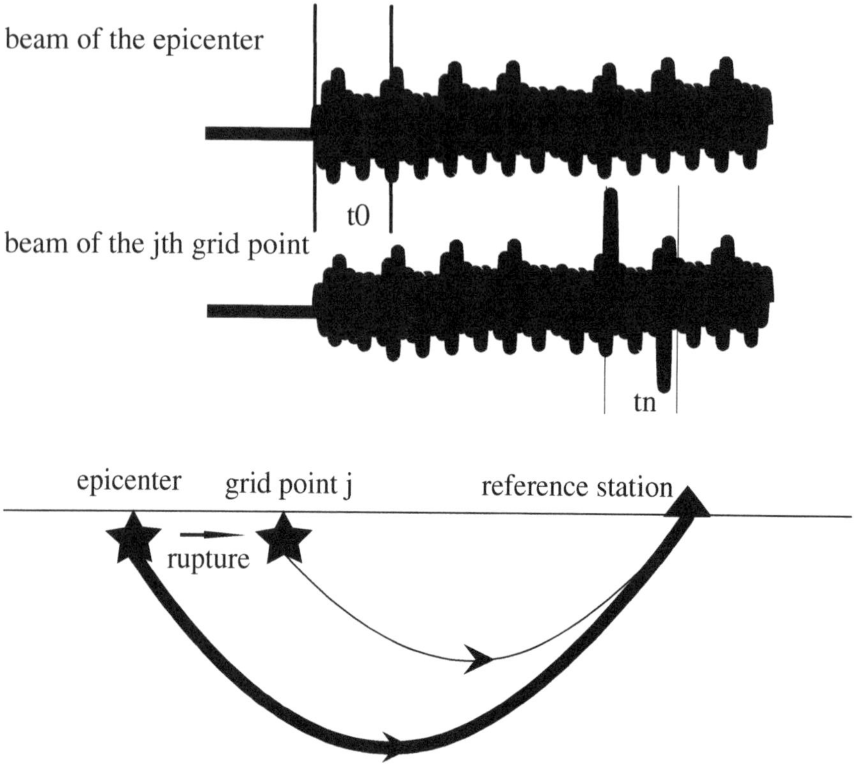

Fig. 3.7 Rupture time of the grid source *j*, obtained by the temporal and spatial relations with the epicenter. The beam of energy of the epicenter between the two thick vertical lines is largest for the first time point t_0. The beam of energy of the *j*th grid point is largest for the time t_n

In this study, a threshold discriminant factor was set to 0.04 because values after 100 s were lower than that value and remained at a low level (Fig. 3.9). Then, rupture fronts with discriminant factors lower than the threshold were removed. As strike-slip mechanisms in the northeastern portions increased, the reflections from free surface, such as *pP* and *sP* phases, would play a more significant role in the seismograms. To correctly derive the rupture time of all fronts from the direct *P* wave, the fronts caused by the reflected phases with longer rupture time must be removed. Most of the fronts jumping in the opposite direction of the rupture are derived from the reflected phases. Thus, removing the fronts that were jumping backward would decrease the effects of the reflected phases significantly. Using these described measures, 29 rupture fronts remaining in Fig. 3.10 are thought to be significant. Their time–distance relationship is given in Fig. 3.11.

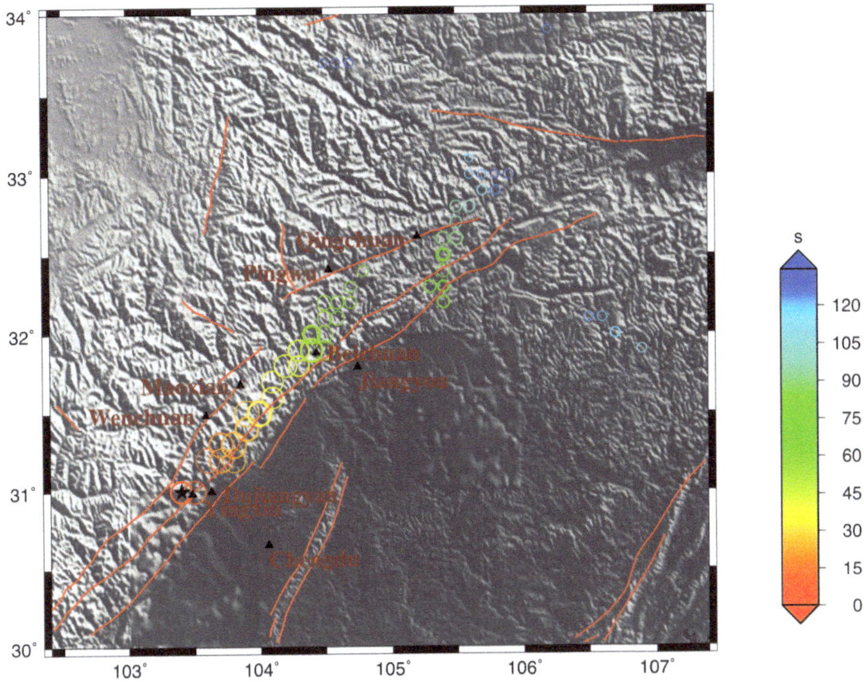

Fig. 3.8 Distribution of high-frequency energy sources before discrimination. Areas of the circles are proportional to the energy; colors are related to their rupture times

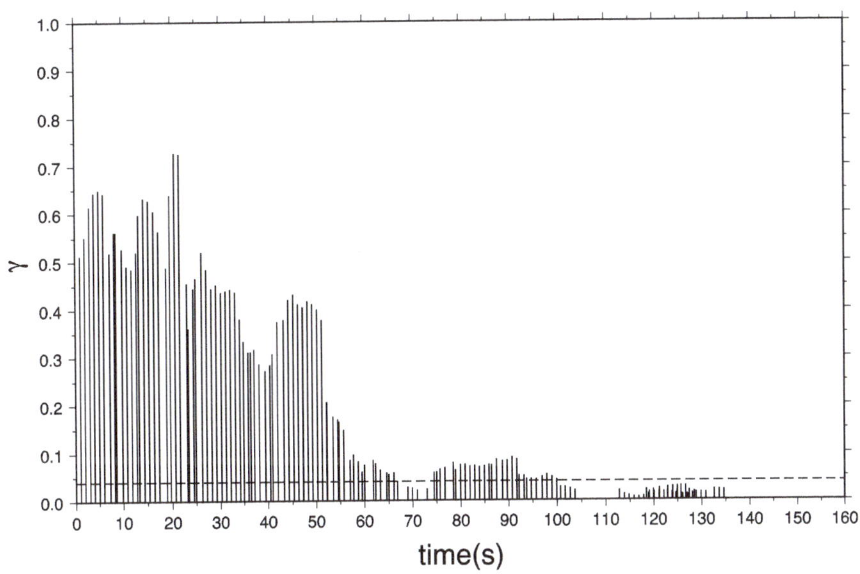

Fig. 3.9 Discriminant factor values of 145 high-frequency energy sources. The dashed line indicates the threshold (0.04). The values are lower than the threshold and remain at a low level after 100 s

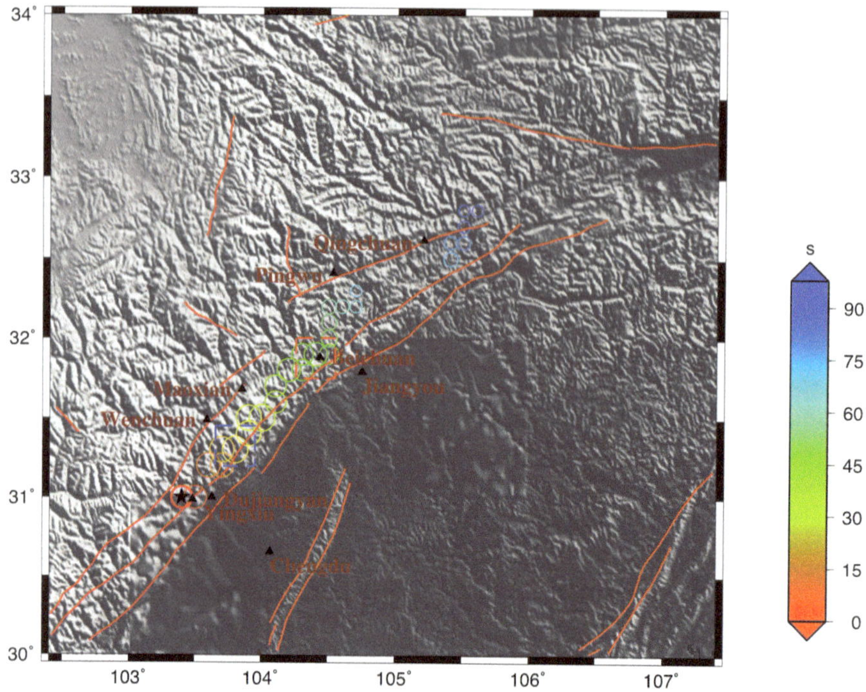

Fig. 3.10 After discrimination, the distribution of high-frequency sources coincides with geometry of the faults (red lines). Blue and red dashed squares represent two major energy-releasing positions during the rupture. The position shown in blue is about 50 km away at about 20 s. The position shown in red one is about 140 km at about 50 s

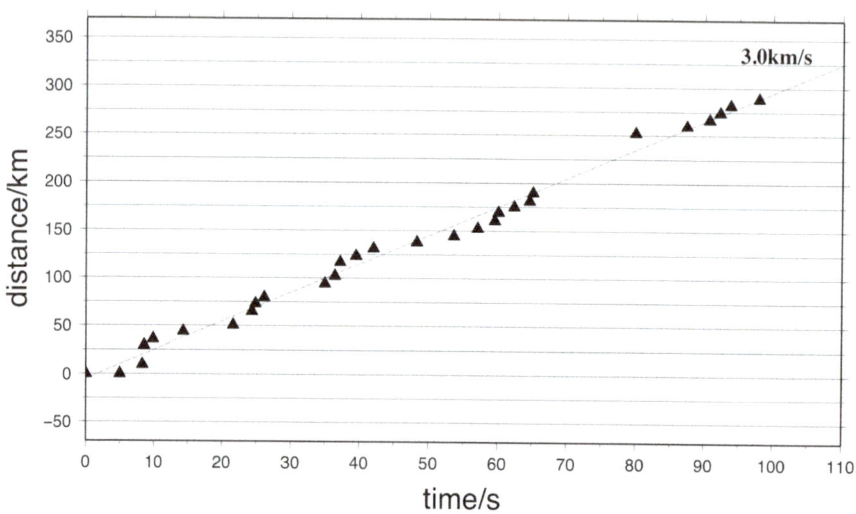

Fig. 3.11 Temporal and spatial relation among effective high-frequency energy sources and epicenter. Dashed line is obtained by linear regression to fit the time–distance relations of the rupture process. Its slope is the mean rupture velocity with a value of 3.0 km/s

3.3 Results

According to the distribution of the rupture fronts given in Fig. 3.10 and their time–distance relationship shown in Fig. 3.11, we can see the total rupture length of the 2008 Wenchuan earthquake is about 290 km, and its duration time is about 98 s. Moreover, the results show that this earthquake mainly unilaterally ruptured northeastward along the Longmen Shan fault zone. The average rupture speed is about 3.0 km/s. There are two major energy-releasing peaks during the rupture period. One is about 50 km away from the epicenter at about 20 s after the earthquake began, and the other is about 140 km northeast to the epicenter near Beichuan county at about 50 s. Furthermore, many rupture fronts with large energy are clustered near these two places. These facts indicate two major subevents occurred near these two locations.

The total duration time of the 2008 Wenchuan earthquake obtained previously is smaller than the result of 110 s obtained using a finite-fault inversion method (Wang et al. 2008). However, it is almost equal to 95 s obtained by using a linear inversion method (Zhang et al. 2009). Because the farthest northeast region of the rupture had weak high-frequency direct P waves, there is a gap of rupture fronts from Pingwu to Qingchuan for the northern portion of the earthquake, which has been presented by other researchers (Zhang 2008). Moreover, many rupture fronts to the northeast endpoint of the Longmen Shan fault zone with energies lower than the threshold of the discriminant factor were removed. Thus, the rupture length is about 290 km, which is consistent with the result derived by inversion methods (Wang et al. 2008; Zhang et al. 2009) and is longer than the Earth surface rupture, which had a length of 240 km (Xu et al. 2008). The analysis of the two major energy-releasing peaks during the rupture period is consistent with the cross section of slip distribution (Wang et al. 2008; Zhang et al. 2009). Because the final results are mostly consistent with those obtained by a variety of inversion methods (as described here) and geological surveys, this method should be an effective way to study the rupture process of the earthquake.

3.4 Discussion

Radiation pattern plays a role in affecting the amplitudes of the direct *P* waves. The radiation pattern may have an effect on the back-projection power distribution. To remove the effect of the radiation pattern, the power of determined rupture fronts could be divided by square of the radiation pattern factor when the focal mechanism of the earthquake is known. Using the focal mechanism of the Wenchuan earthquake derived by Zhang et al. (2009), the high-frequency sources evolving with time was obtained as shown in Fig. 3.12. The imaging results show that two high-frequency releasing peaks are slightly spatially deviated from those in

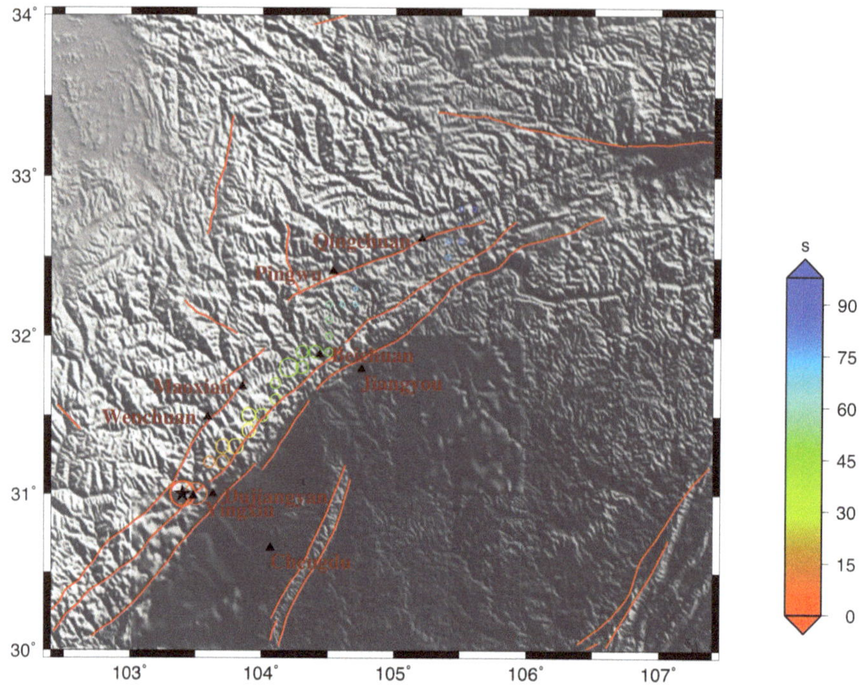

Fig. 3.12 Results with corrections for the effects of radiation pattern. The relative energies of all high-frequency sources are shown

Fig. 3.10. The first one is moved from 50 km away from the epicenter to the epicenter nearby Yingxiu town, and the second one is moved southwestward, nearby Beichuan county. This agrees well with the most damaging zone on the surface (Mori and Smith 2009). However, the focal mechanism of a large earthquake is not immediately available. Moreover, the complexity of the rupture of large earthquakes makes it difficult to do the radiation pattern calibration. Thus, in a fast fashion, the imaging results as shown in Fig. 3.10 are acceptable.

For the shallow rupture, the surface-reflected waves such as *pP* and *sP* can significantly interfere the imaging from the direct *P* waves. Especially for the 2008 Wenchuan earthquake, the focal mechanism starts with thrust and gradually transits into thrust plus large component of dextral strike-slip (Shen et al. 2009). This changing of focal mechanism may promote the interference of the surface-reflected phases.

To assess the influence of the two depth phases, the amplitude ratios of the *pP* and *sP* to the direct *P* at the Australian seismic network during the last 60 s are estimated on the basis of the kinematic rupture model by Zhang et al. (2009). The amplitude of *pP* is estimated by the product of radiation pattern factor of the up-going *p* wave and the reflectivity of *P* at the surface. Similarly, the amplitude of

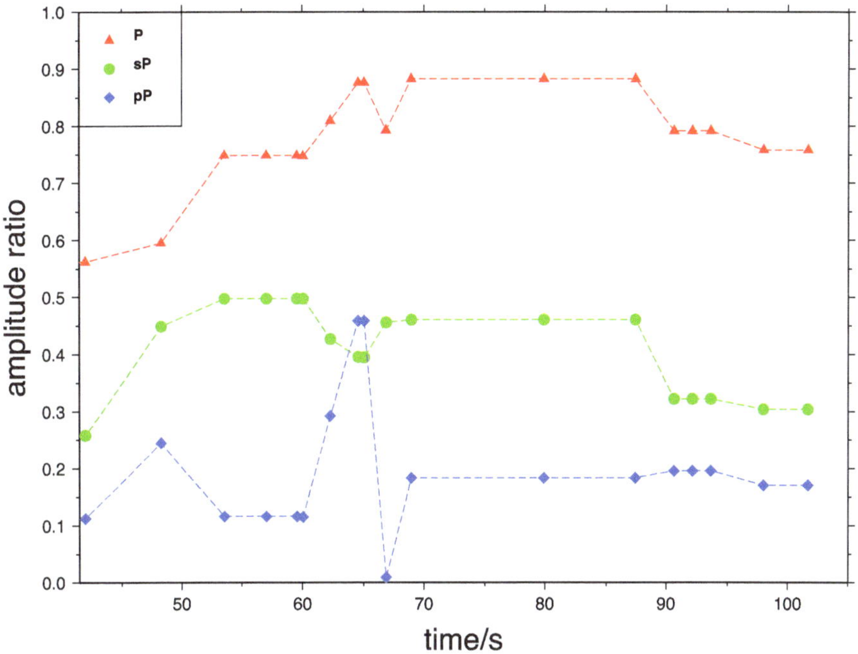

Fig. 3.13 Amplitude ratios of *P*, *sP*, and *pP* 42 s after the origin time of the Wenchuan earthquake. *P*, *sP*, and *pP* are indicated by red triangles, green dots, and blue diamonds, respectively

sP is constrained by the product of radiation pattern factor of up-going *s* wave and the reflectivity of *P* at the surface as well as the velocity ratio of *P* and *S* waves $\sqrt{3}$. The resultant amplitude ratios (Fig. 3.13) show that the amplitudes of *P* waves are much larger than those of the depth phases. This indicates that the whole 160 s *P* waves can be treated as the direct *P* waves without involving in significant artifacts from the depth phases. Moreover, as described above, the removal of imaging subevents jumping back can also reduce the imaging from the depth phases. The two evidence lines support the reasonability of the imaging results derived from the direct *P* waves.

3.5 Conclusions

In this study, a non-plane-wave assumption-based array technique, the relative back-projection method, was introduced in detail and applied to track the rupture process of the May 12, 2008, Wenchuan earthquake. The imaging results show that the earthquake unilaterally ruptured northeastward over a length of 290 km during ~100 s. The fast rupture on the Longmen Shan fault caused great damaging to two regions (Yingxiu town and Beichuan county), where the two high-frequency releasing peaks were located; this may indicate that the back-projection results can be used as an indicator of the damaging zones nearby the causative faults of the earthquakes.

Compared with the source inversion method, which requires setting of the initial model with rupture velocity and fault parameters and involves inversion iterations, we need to know only the epicenter in our method to perform beamforming. Thus, this method is a handy tool to perform rapid estimation of the rupture process for large earthquakes.

References

Birtill, J.W., and F.E. Whiteway. 1965. The application of phased arrays to the analysis of seismic body waves. *Philosophical Transactions of the Royal Society of London Science A* 258: 421–493.

Burchfiel, B.C., L.H. Royden, R.D. van der Hilst, et al. 2008. A geological and geophysical context for the Wenchuan earthquake of 12 May 2008, Sichuan, People's Republic of China. *GSA Today* 18 (7): 4–11.

Capon, J. 1973. Signal processing and frequency-wavenumber spectrum analysis for a large aperture seismic array. *Method in Computational Physics* 13: 1–59.

Davies, D., E.J. Kelly, and J.R. Filson. 1971. Vespa process for analysis of seismic signals. Nature Physical Science 232: 8–13.

Douglas, A. 1998. Making the most of the recordings from short-period seismometer arrays. *Bulletin of the Seismological Society of America* 88 (5): 1155–1170.

Du, H.L., L.S. Xu, and Y.T. Chen. 2009. Rupture process of the 2008 great Wenchuan earthquake from the analysis of the Alaska array data. *Chinese Journal of Geophysics (in Chinese)* 52 (2): 372–378.

Ge, Z.X., and X.F. Chen. 2008. Point source stacking method to compute synthetic seismogram of finite moving planar source. *Acta Scientiarum Naturalium Universitatis Pekinensis. Beijing University (Natural Science)* 44: 407–412.

Harjes, H.-P., and M. Henger. 1973. Array-Seismologie. *Zeitschrift Geophysics* 39: 865–905. (in German).

Ishii, M., P.M. Shearer, H. Houston, and J.E. Vidale. 2005. Extent, duration and speed of the 2004 Sumatra-Andaman earthquake imaged by the Hi-Net array. *Nature* 435 (7044): 933–936.

Kruger, F., and M. Ohrnberger. 2005. Tracking the rupture of the Mw & equals; 9.3 Sumatra earthquake over 1150 km at teleseismic distance. *Nature* 435 (7044): 937–939.

Lay, T. 1987. Analysis of near source contributions to early P-wave coda for underground explosions, III, Inversion for isotropic scatterers. Bulletin of the Seismological Society of America 77: 1767–1783.

Lynnes, C.S., and T. Lay. 1989. Inversion of P coda for isotropic scatters at the Yucca Flat test site. *Bulletin of the Seismological Society of America* 79: 790–804.

Mori, J., C. Smyth. 2009. A summary of seismological observations for the 2008 Wenchuan, China earthquake, Investigation Report of the 2008, Wenchuan Earthquake, China, Grant-in-Aid for Special Purpose of 2008, Ministry of Education, Culture, Sports, Science and Technology (MEXT), Japan, No. 20900002, 213–214.

Ni, S.D., H. Kanamori, and D. Helmberger. 2005. Energy radiation from the Sumatra earthquake. *Nature* 434: 582.

Posmentier, E.S., and R.W. Herrmann. 1971. Cophase, an ad hoc array processor. *Journal Geophysical Research* 76: 2194–2201.

Rost, S., and C. Thomas. 2002. Array seismology: Methods and applications. *Reviews of Geophysics* 40 (3): 1008.

Schweitzer J., J. Fyen, S. Mykkeltveit. 2002. Seismic arrays. Chapter 9 in New Manual of Seismological Observatory Practice. 1–51.

Shen, Z.-K., J. Sun, P. Zhang, et al. 2009. Slip maxima at fault junctions and rupturing of barriers during the 2008 Wenchuan earthquake. *Nature Geoscience* 2 (10): 718–724.

Walker, K.T., M. Ishii, and P.M. Shearer. 2005. Rupture details of the 28 March 2005 Sumatra Mw 8.6 earthquake imaged with teleseismic P waves. *Geophysical Research Letters* 32: L24303.

Wang, W.M., L.F. Zhao, J. Li, and Z.X. Yao. 2008. Rupture process of the Ms 8.0 Wenchuan earthquake of Sichuan, China, Chinese Journal of Geophysics (in Chinese) 51(5): 1403–1410.

Xu, X.W., X.Z. Wen, J.Q. Ye et al. 2008. The Ms 8.0 Wenchuan earthquake surface ruptures and its seismogenic structure. Seismology and Geology (in Chinese) 30(3): 597–629.

Xu Y., K.D. Koper, O. Sufri, et al. 2009. Rupture imaging of the Mw 7.9 12 May 2008 Wenchuan earthquake from back projection of teleseismic P waves. Geochemistry, Geophysics, Geosystems 10(4): 1–17.

Zhang, P.Z., X.W. Xu, X.Z. Wen, and Y.K. Ran. 2008. Slip rates and recurrence intervals of the Longmen Shan active fault zone, and tectonic implications for the mechanism of the May 12 Wenchuan earthquake, 2008, Sichuan, China. *Chinese Journal of Geophysics (in Chinese)* 51 (4): 1066–1073.

Zhang, Y. 2008. Study on the inversion methods of source rupture process, Ph.D. Thesis, Peking University.

Zhang, Y., W.P. Feng, L.S. Xu, C.H. Zhou, and Y.T. Chen. 2009. Spatio-temporal rupture process of the 2008 great Wenchuan earthquake. *Science in China Series D: Earth Sciences* 52 (2): 145–154.

Chapter 4
Back-Projection Imaging the 2010 Mw 8.8 Chilean Earthquake

4.1 Introduction

On February 27, 2017, at 06:34:00 (UTC) an Mw 8.8 megathrust earthquake struck nearby Concepcion, South Chile, where the Nazca plate is subducting beneath the overriding South American plate. The earthquake was located at (35.905° S, 72.733° W) at a focal depth of 35 km (provided by USGS).

Soon after the earthquake, several groups of researchers have carried out source studies on the earthquake. Liu et al. (2010) and Kiser and Ishii (2011) applied the back-projection method to investigate the earthquake. The back-projection method requires very few presetting parameters such as the epicenter and a velocity model like IASPEI91 (Kennett and Engdahl 1991) using vertical-component *P* waves. Since only slant-stacking technique is used in the back-projection, the rupture of the earthquake can be resolved quickly. However, Liu et al. (2010) chose a very long time window (60 s); this led to a low spatial resolution in imaging results. Typically, the rupture of the earthquake was imaged on a horizontal plane, though it is suitable for gently-dipping faults.

Here, we applied an improved back-projection method (Zhang and Ge 2010; Zhang et al. 2011) to study the 2010 Chilean earthquake using three-component velocity seismograms at the USArray. To resolve the rupture in detail, a fault plane constrained by USGS was used as the source region, where the waveforms were back-projected. Moreover, the fault plane was divided into northern and southern segments, on which the waveforms were back-projected separately to resolve the bilateral rupture of the earthquake.

H. Zhang, *Imaging the Rupture Processes of Earthquakes Using the Relative Back-Projection Method*, Springer Theses,
https://doi.org/10.1007/978-3-662-55239-1_4

4.2 Data and Methodology

Soon after the 2010 Chilean earthquake, 415 three-component seismograms at the USArray transportable network were downloaded from the IRIS DMC (http://www.iris.edu/dms/wilber.htm, last accessed on May 11, 2010). Considering the significance of the signal-to-noise ratio to imaging, 309 three-component direct *P* waves were eventually kept with a station distribution as shown in Fig. 4.1. The regional array composed of these stations has an aperture of $\sim$3000 km and is 68°–92° away from the epicenter. The reference station Q26A is located at the center of the array with an epidistance of 80°.

These teleseismic *P* waves require preprocessing before applied to back-projection imaging. Due to the influence of temperature and atmospheric pressure, seismograms at some stations could have a trend or daily fluctuation. The dominant frequency spectrum mainly focuses on a range of 0.01–1.0 Hz with a peak at 0.08 Hz (Fig. 4.2). The higher the data used are, the higher the spatial resolution is. The teleseismic high-frequency content mainly stems from the extension of the rupture front (Madariaga 1977, 1983). Moreover, reflection phases such as *PP* and

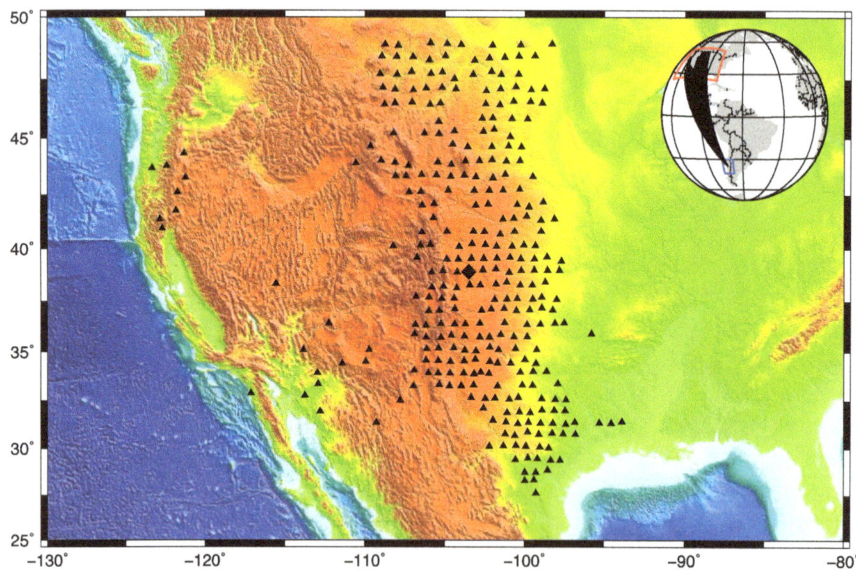

Fig. 4.1 Station distribution of the USArray transportable array (TA, black triangles). The black diamond denotes the reference station Q26A among 309 stations. Inset, the TA is located inside the red frame, and the 2010 Chilean earthquake occurred in the blue frame. The black curves represent the ray paths

PcP have longer ray paths in the high-attenuation zones in earth's interior; this leads to less high-frequency content included in the reflection waves relative to the direct *P* waves (Ni et al. 2005). However, the high-frequency signals from the same source may significantly differ for stations a certain distance apart. Similarity is a key factor for the back-projection imaging. Thus, the *P* waves used in this study were filtered in a frequency band of 0.4–2.0 Hz. The travel time difference between the phase *PP* and direct *P* waves to the reference time is 183.4 s. To preclude the inference from *PP*, the time window was chosen to 180 s after the first *P* arrival.

To carry out the relative back-projection imaging of the 2010 Chilean earthquake, a fault plane with the hypocenter at the center was set to 700 $\times$ 250 km with a strike of 18° and a dip of 18° according to the source parameters provided by the USGS. The fault plane was gridded into 7191 blocks with spacing of 5 km both at the strike and at the dip. The rupture of the earthquake has been observed to propagate bilaterally (Lay et al. 2010). To resolve the bilateral earthquake, the fault plane was separated into northern and southern segments. Next, we conducted the relative back-projection imaging on the two fault segments.

Some of the resultant imaging subevents may be false due to the interference of the depth phases (Zhang and Ge 2010). To remove these artifacts, a discriminant criteria proposed by Du et al. (2009) was applied to the resultant 320 subevents as shown in Fig. 4.3a. After culling these artifacts with a threshold of 0.05, the northern segment has 41 subevents, and the southern segment has 13 subevents as shown in Fig. 4.3b.

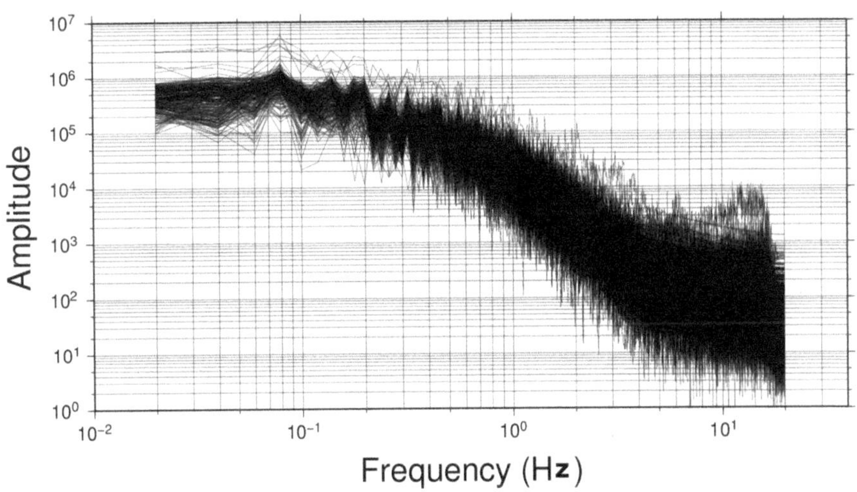

Fig. 4.2 The amplitude spectrum of the vertical-component seismograms at the TA. The amplitude spectrum dramatically decreases when the frequency is higher than 0.1 Hz

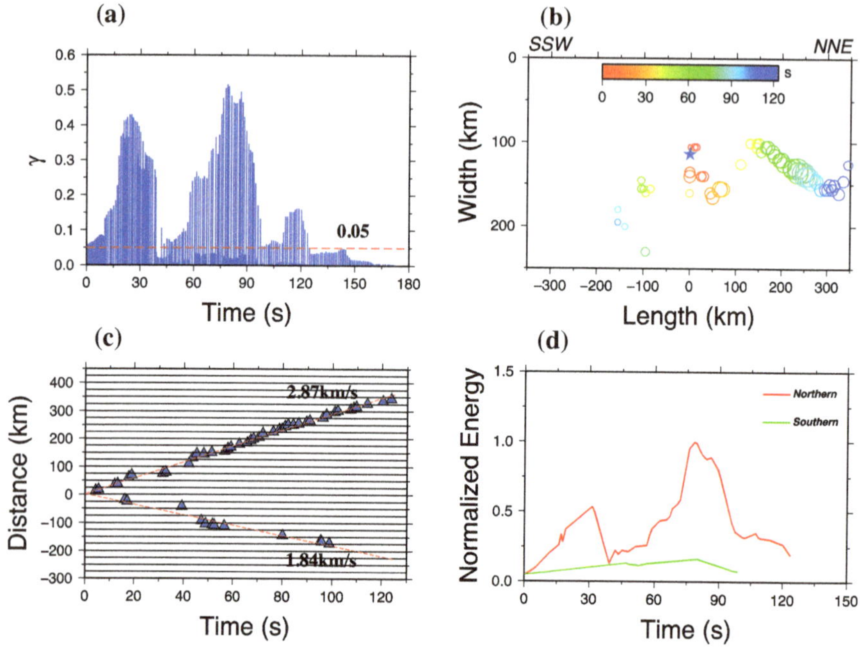

Fig. 4.3 The relative back-projection imaging of the 2010 Chilean earthquake: **a** Discriminator of the rupture front with a threshold of 0.05. There are two peaks at 25 s and 80 s, respectively; **b** The subevents (circles) evolving with time and distributing on the fault. The areas of the circles are proportional to the energy of the subevents, and the colors indicate the rupture time; **c** The distance of the subevent as a function of rupture time; and **d** The normalized power of the subevent against the rupture time for the northern and southern branches of the rupture

4.3 Results

The imaging subevents projected at the surface as shown in Fig. 4.3 show that the 2010 Mw 8.8 Chilean earthquake ruptured simultaneously northward and southward during 125 s. Moreover, the northern branch of the rupture dominates the high-frequency energy releasing with a rupture length of 340 km, which agrees well with the aftershocks in 24 h after the quake of the earthquake. Since the southern rupture is relatively weak, the subevents after 100 s are not resolved by the back-projection imaging. But the rupture length of the southern branch is at least 173 km as shown in Fig. 4.3c. Thus, the bilateral Chilean earthquake ruptures over an overall length of ~513 km. On the fault plane, the rupture extends 100 km along the dip (Fig. 4.3b).

The average rupture velocity of the northern branch is ~2.87 km/s and that of the southern branch is ~1.84 km/s (Fig. 4.3c). For the northern branch, the average rupture velocity is consistent with that of 2.8–3.0 km/s estimated by Vigny et al. (2011). There are two high-frequency energy releasing peaks: one 75 km away from the hypocenter at 16 s and one 230 km away from the hypocenter at 79 s (Fig. 4.3d). For the southern branch, only one high-frequency energy releasing peak is located 145 km away at 80 s.

In between latitude 34.7 and 35.4° S, there exist two subevents during the period of 19–42 s (Fig. 4.4). The rupture in this period could be a transition from the initial

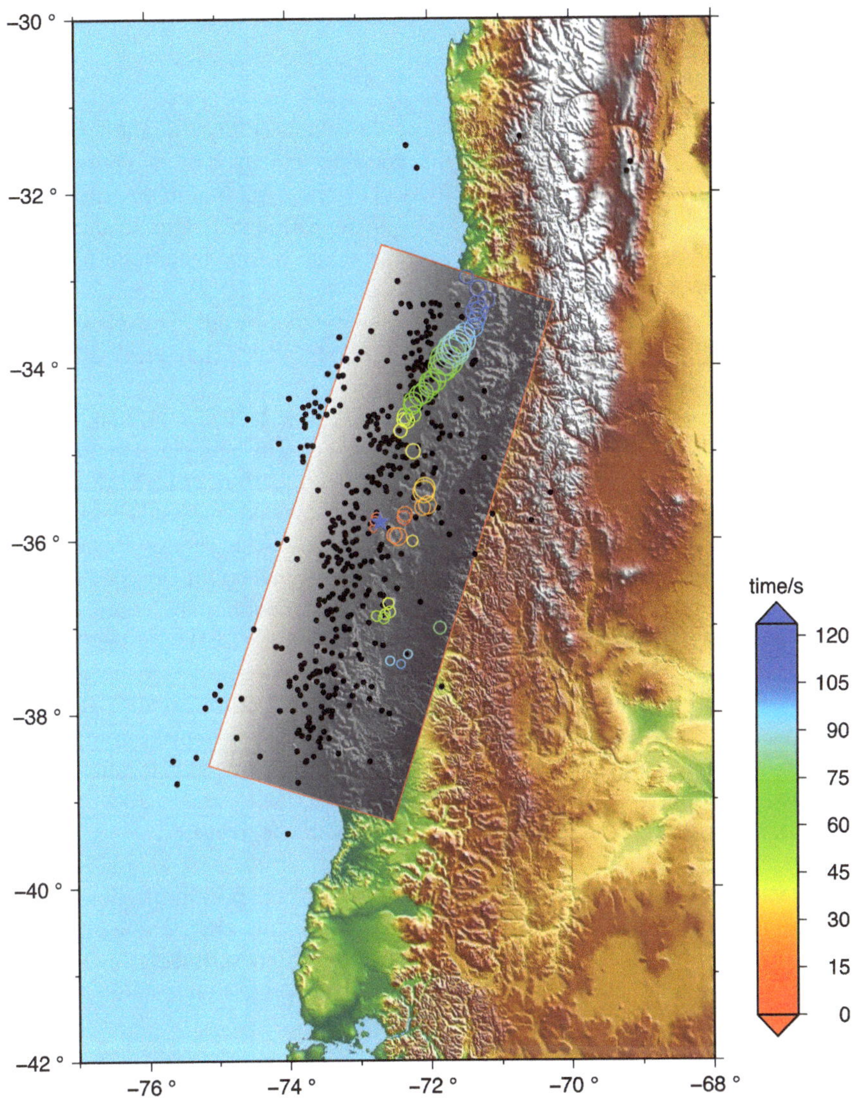

Fig. 4.4 The surface projection of the back-projection imaging of the rupture. The gray-scaled region is the surface projection of the fault plane. The circles indicate the subevents with area proportional to the power and color representing the rupture time. Black dots denote the aftershocks in 24 h, and the blue star depicts the epicenter of the earthquake

rupture deeper than the hypocenter to another northward rupture segment relatively shallower. Additionally, the southern subevents are located deeper relative to the northern branch. Combining the observation of the rupture velocity, it seems like the deeper rupture has lower rupture velocity. This discovery has also been observed in the 2015 Nepal earthquake (Zhang et al. 2016).

4.4 Discussion

The dominate rupture of the 2010 Chilean earthquake is northward. There are much more subevents in the northern branch that could be explained by two possibilities. The first possibility is the rupture directivity. The array we used in this study is the USArray transportable array, located north of the earthquake. Due to the rupture directivity, the seismic signals from the northern branch have more high-frequency content than those from the southern branch (Zhang et al. 2012). Alternatively, the dominate rupture of the 2010 Chilean earthquake is northward. This possibility is confirmed by the source studies using seismic body waves (Lay et al. 2010) and geodetic data (Tong et al. 2010).

The sparse subevents are discovered in between the latitudes 34.7 and 35.4°S. This could be caused by the smoothing or rare rupture. The fault plane in this region could have barriers or aseismic zones. Moreno et al. (2010) discovered this region, just located in between two large locked areas, has few deformation. This indicates this region should be an aseismic zone, radiating little seismic energy. However, the fault segment north of the aseismic zone could be triggered by the stress transferring from the preceding rupture; this makes the rupture continuing northward.

Intriguingly, the two high-frequency energy releasing peaks of the northern branch is separated by the aseismic zone (Fig. 4.4). The two peaks are located at the edges of the two areas having the maximum slip deficits (Moreno et al. 2010). Kiser and Ishii (2011) also explained the northern rupture of the earthquake composed of the two subevents separated by the aseismic zone. The deeper location of the high-frequency subevents than the totally coupled fault areas agrees with the frequency-dependent rupture features discovered in the megathrust earthquakes (Lay et al. 2012).

Compared with the horizontal fault used in most of the back-projection studies, setting a given fault plane is more realistic. Moreover, the separate imaging of the northern and southern segments facilitates resolving the subevents in the weak southern branch. Thus, these modifications of the back-projection imaging make the source analysis on the earthquake more accurate.

4.5 Conclusions

The 2010 Mw 8.8 Chilean earthquake was investigated by back-projecting three-component teleseismic *P* waves to the source region using the relative back-projection method. The rupture of the earthquake is roughly composed of three segments over a fault plane of 515 $\times$ 100 km. The rupture commences at the epicenter and ruptures simultaneously northward and southward. The northern branch significantly dominates the rupture of the earthquake. The northern branch of the rupture has a length of 340 km over duration of 125 s at a speed of $\sim$2.87 km/s. Meanwhile, the southern branch ruptures over a length of 173 km/s in duration of $\sim$100/s at a speed of $\sim$1.84 km/s. In the northern branch, there are two subevents, which correspond to the two high-frequency energy releasing peaks. The two subevents are separated by an aseismic zone in between the two subevents.

References

Du, H.L., L.S. Xu, and Y.T. Chen. 2009. Rupture process of the 2008 great Wenchuan earthquake from the analysis of the Alaska array data. *Chinese Journal of Geophysics (in Chinese)* 52 (2): 372–378.

Kennett, B.L.N., and E.R. Engdahl. 1991. Traveltimes for global earthquake location and phase identification. *Geophysical Journal International* 105 (2): 429–465.

Kiser, E., and M. Ishii. 2011. The 2010 Mw 8.8 Chile earthquake: Triggering on multiple segments and frequency-dependent rupture behavior. *Geophysical Research Letters* 38 (7).

Lay, T., C.J. Ammon, H. Kanamori, et al. 2010. Teleseismic inversion for rupture process of the 27 February 2010 Chile (Mw 8.8) earthquake. *Geophysical Research Letters* 37: L13301.

Lay, T., H. Kanamori, C.J. Ammon, et al. 2012. Depth-varying rupture properties of subduction zone megathrust faults. Journal of Geophysical Research: Solid Earth (1978–2012) 117 (B04311).

Liu, N., F.L., Niu, Q.F. Chen, and Y. Chen. 2010. Imaging the rupture of the 2010 Mw 8.8 Chile earthquake with a broadband seismic array. *Chinese Journal Geophysics* 53(7): 1605–1610. doi:https://doi.org/10.3969/j.issn.0001-5733.2010.07.011.

Madariaga, R. 1977. High-frequency radiation from crack (stress drop) models of earth-quake faulting. *Geophysical Journal of the Royal Astronomical Society* 51 (3): 625–651.

Madariaga, R. 1983. High frequency radiation from dynamic earthquake fault models. *Annales Geophysicae* 1 (1): 17–23.

Moreno, M., M. Rosenau, and O. Oncken. 2010. 2010 Maule earthquake slip correlates with pre-seismic locking of Andean subduction zone. *Nature* 467 (7312): 198–202.

Ni, S.D., H. Kanamori, and D. Helmberger. 2005. Energy radiation from the Sumatra earthquake. *Nature* 434: 582.

Tong, X., D. Sandwell, K. Luttrell, et al. 2010. The 2010 Maule, Chile earthquake: Downdip rupture limit revealed by space geodesy. *Geophysical Research Letters* 37: L24311.

Vigny, C., A. Socquet, S. Peyrat, J.- C. Ruegg, M. Metois, R. Madariaga, S. Morvan, M. Lancieri, R. Lacassin, J. Campos, D. Carrizo, M. Bejar-Pizarro, S. Barrientos, R. Armijo, C. Aranda, M.-C. Valderas-Bermejo, I. Ortega, F. Bondoux, S. Baize, H. Lyon-Caen, A. Pavez, J. P. Vilotte, M. Bevis, B. Brooks, R. Smalley, H. Parra, J.- C. Baez, M. Blanco, S. Cimbaro, and E. Kendrick. 2011. The 2010 Mw 8.8 Maule Megathrust Earthquake of Central Chile, Monitored by GPS. *Science* 332 (6036): 1417–1421.

Zhang, H., and Z. Ge. 2010. Tracking the rupture of the 2008 Wenchuan earthquake by using the relative back-projection method. *Bulletin of the Seismological Society of America* 100 (5B): 2551–2560.

Zhang, H., Z. Ge, and L. Ding. 2011. Three sub-events composing the 2011 off the Pacific coast of Tohoku Earthquake (Mw 9.0) inferred from rupture imaging by back-projecting teleseismic P waves. *Earth, Planets and Space* 63 (7): 595–598.

Zhang, H., J. Chen, and Z. Ge. 2012. Multi-fault rupture and successive triggering during the 2012 Mw 8.6 Sumatra offshore earthquake. *Geophysical Research Letters* 39: L22305.

Zhang, H., S. van der Lee, and Z. Ge. 2016. Multiarray rupture imaging of the devastating 2015 Gorkha, Nepal, earthquake sequence. *Geophysical Research Letters* 43: 584–591. https://doi.org/10.1002/2015GL066657.

Chapter 5
Multi-segment, Frequency-Dependent Rupture of the 2011 Mw 9.0 Tohoku Earthquake

5.1 Introduction

On March 11, 2011, the magnitude 9.0 Tohoku earthquake struck off the northeast coast of Honshu, Japan [UTC 05:46:23, hypocenter 38.322° N, 142.369° E, 32 km depth (US Geological Survey (USGS), http://earthquake.usgs.gov/earthquakes/eqinthenews/2011/usc0001xgp/)]. The disastrous earthquake and the accompanying powerful tsunami led to tens of thousands of deaths and many missing. This disaster emphasizes the importance of rapidly and accurately imaging the rupture process for post-earthquake emergency response and tsunami warnings.

In recent years, the back-projection method has been used as an effective approach to image the spatial-temporal rupture process (Ishii et al. 2005; Kruger and Ohrnberger 2005; Walker et al. 2005). The only prior parameter used in this method is the location of the hypocenter, which was determined in almost real time by the USGS. Meanwhile, the calculations of the back-projection method are so straightforward that it takes no more than half an hour to effectively image the rupture process. To image the two-dimensional spatial rupture process of the 2011 Tohoku earthquake, we apply here a two-step back-projection method (Zhang and Ge 2010) to process the broadband teleseismic *P* waves that were recorded by 151 seismic stations of the European seismic network, referred to as the EU.

5.2 Data

We downloaded teleseismic *P* wave data of the 2011 Tohoku earthquake recorded by 344 European stations from the website (http://www.orfeus-eu.org/cgi-bin/wilberII/wilberpage1.pl) of the Observations and Research Facilities for European Seismology (ORFEUS). Considering the epicentral distance and azimuth coverage of stations, as well as the quality of their data, 151 broadband seismograms from

H. Zhang, *Imaging the Rupture Processes of Earthquakes Using the Relative Back-Projection Method*, Springer Theses,
https://doi.org/10.1007/978-3-662-55239-1_5

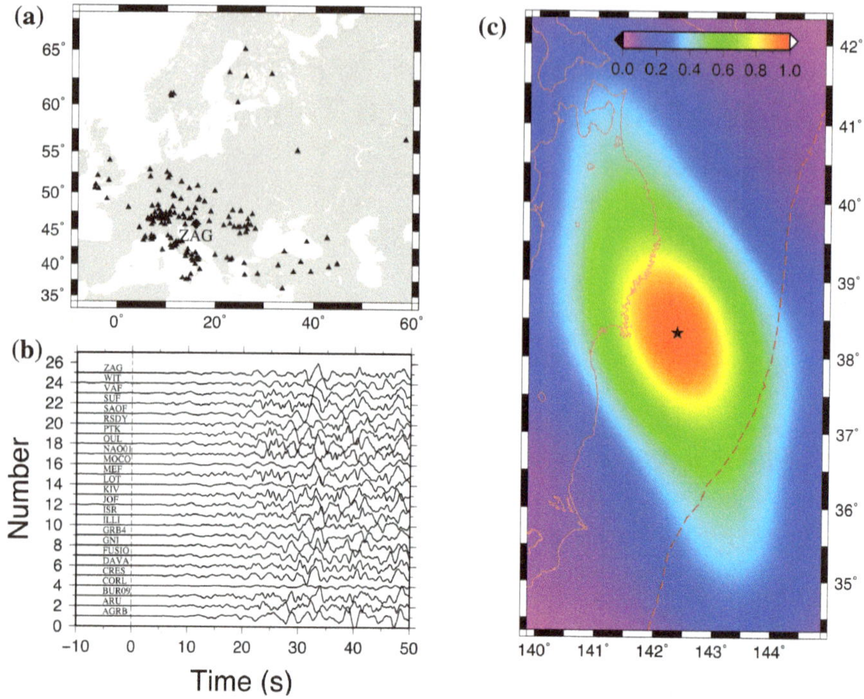

Fig. 5.1 Features related to the stations of the European network. **a** Distribution of 151 stations (black triangles and a black diamond) throughout Europe that compose the Euro network. The large black diamond with the label "ZAG" indicates the reference station. **b** Examples of the vertical-component high-quality normalized seismograms from 25 stations with a time length of 60 s, which are high-pass filtered with a cutoff frequency of 0.2 Hz. **c** The array response function of the Euro network to the source region at 0.2 Hz. The black star indicates the epicenter of the Tohoku earthquake. The dashed line depicts the Japan Trench. The contour lines indicate the normalized beam energy with an interval of 0.1

the stations were selected as shown in Fig. 5.1a. The reference station (ZAG in Fig. 5.1a), with the shortest distance to the center (47.654° N, 14.912° E) of the Euro network, is about 82.5° away from the epicenter and has an azimuthal angle of ~325°. The orientation of the Euro network to the epicenter is nearly perpendicular to the strike direction (193°, presented by USGS). This orientation, together with the large aperture, of the Euro network is of great utility in resolving the rupture along the strike direction.

Teleseismic *P* waves over a period of time of 220 s were used to isolate the *P* wave from other phases, such as *PP*. To save calculation time, only vertical component seismograms were considered; i.e., only the reference (ZAG) and 24 other traces shown in Fig. 5.1b. Considering the coherence and resolution of the data, seismograms with a sample rate of 20 Hz were high-pass filtered with a cutoff frequency of 0.2 Hz, at which the array response function (Rost and Thomas 2002;

Xu et al. 2009) was calculated to illuminate the array geometry and azimuth on the spatial resolution (Fig. 5.1c).

5.3 Methodology

This study employed the two-step back-projection method described by Zhang and Ge (2010). At the first step, *P* waves were back-projected to find locations of the rupture fronts at the apparent rupture times. Then, rupture times of the fronts were derived from their own locations and apparent rupture times.

To identify the rupture fronts of the 2011 Tohoku earthquake, the source region (34.32°–42.32° N, 139.87°–144.87° E) was spatially gridded into blocks of 0.1° × 0.1°. This process resulted in 4131 grid points referring to potential rupture fronts. Due to the small dip of the fault plane (14°, given by USGS), the potential rupture fronts were fixed at a depth (Ishii et al. 2007) of 32 km, the hypocenter depth given by USGS. The theoretical travel time was calculated in one-dimensional earth velocity model AK135 (Kennett et al. 1995) for each pair of potential rupture front and station. However, due to lateral heterogeneities along ray paths, there is always a small difference between theoretical and real travel times (Fig. 5.1b). Therefore, the theoretical travel times from the stations to the hypocenter were calibrated by aligning the initial 30 s of the teleseismic *P* waves through cross-correlation. These time calibrations of the hypocenter were applied to other potential fronts (Xu et al. 2009).

5.4 Results and Analyses

After the back-projection imaging, the imaging power distribution of the 2011 Tohoku earthquake evolving with time was derived as shown in Fig. 5.2. The subevents at 0 and 15 s were located at the epicenter. On basis of the tests in Chap. 2, the earthquake did not rupture along the strike. Since the EU has low resolution along the dipping direction of the fault, the rupture could propagate along the dip. At 31.2 s, the rupture front was located northeast of the epicenter, was imaged north of the epicenter at 46.7 s, and was imaged northwest of the epicenter with a large high-frequency burst at 61.5 s. However, the rupture came back at the epicenter at 74.9 s. There are two possibilities for the observation. One is that the rupture commenced at the epicenter after the end of the northern branch of the rupture. The other one is that the northern and southern branches were ruptured simultaneous. The centroid of the two subevents was located nearby the epicenter and thus imaged. At this moment, the northern branch of the rupture was far away from the epicenter; this required the southern rupture far away from the epicenter as well. However, the afterward subevents on the southern branch were still located close to the epicenter and continued to extend southward, supporting the first

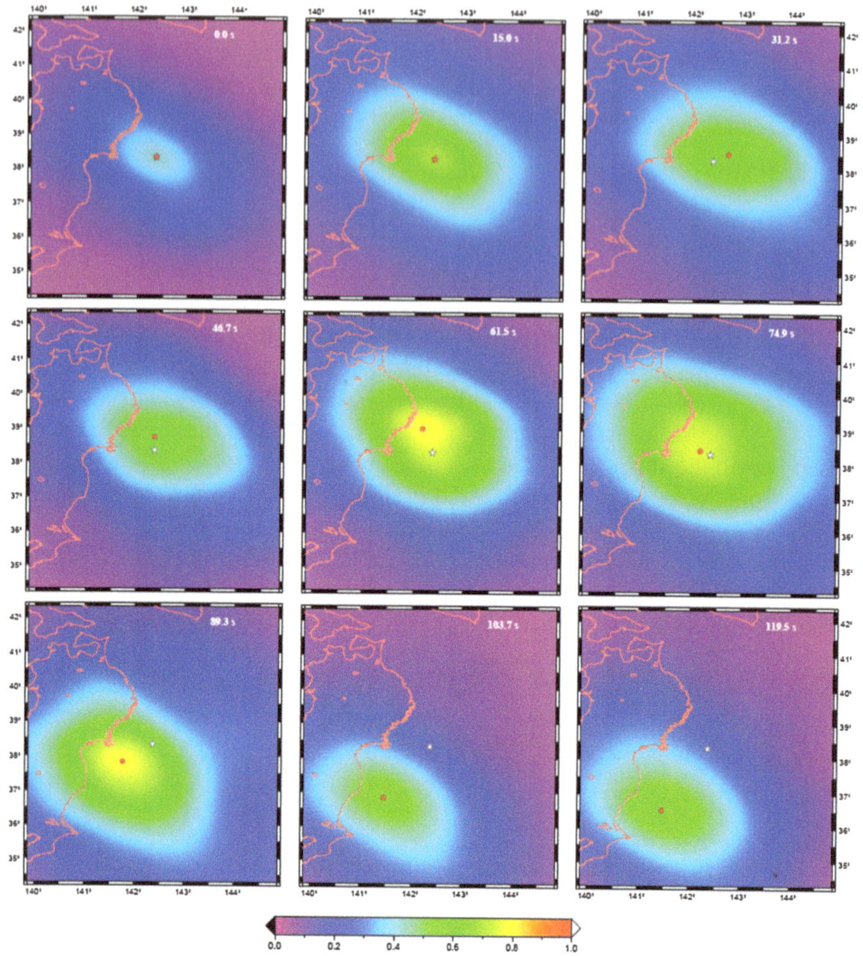

Fig. 5.2 Rupture fronts at 0, 15, 31.2, 46.7, 61.5, 74.9, 89.3, 103.7, and 119.5 s. The energy of all the grids is normalized by the maximum energy during the whole rupture process

possibility. Thus, the Tohoku earthquake initiated the southward rupture at the epicenter at 75 s, and then, the rupture extended southward afterwards as shown in the snapshots at 89.3, 103.7, and 119.5 s (Fig. 5.2).

Projecting the back-projection imaging subevents on the surface, the rupture process of the earthquake was obtained as shown in Fig. 5.3a. The image was used to generate the normalized energy and rupture time relationships of the rupture fronts as shown in Fig. 5.3b. The rupture ending time was determined as 143 s due to the sudden decrease in the energy released (Fig. 5.3b). To mitigate the smearing of the fronts in time, rupture fronts located at the same place for several continuous time steps were removed except one with the maximum releasing energy. The image of the rupture process was obtained as shown in Fig. 5.3a. The rupture

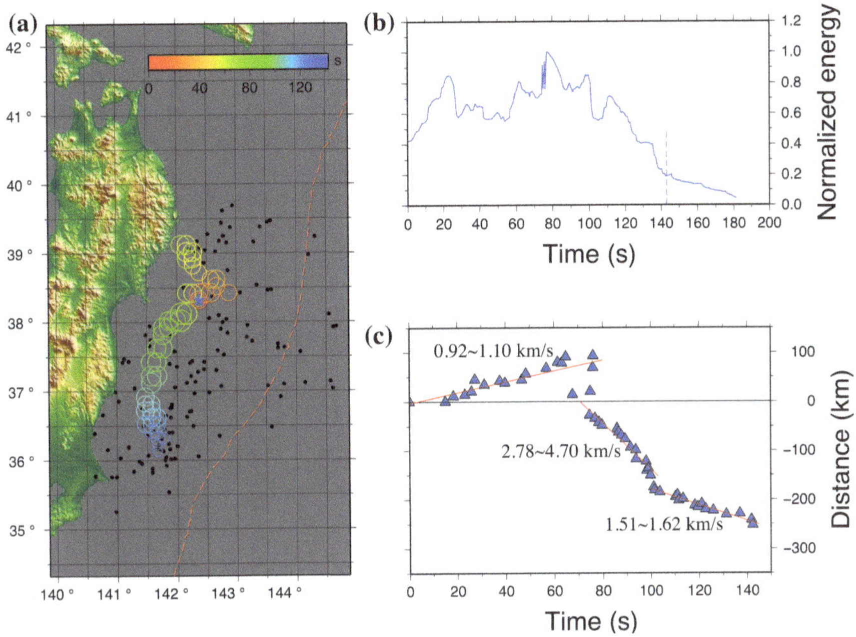

Fig. 5.3 Rupture imaging of the 2011 Tohoku earthquake. **a** Spatial and temporal (color bar shows the rupture time) distribution of three high-frequency energy sources during the whole rupture time of at least 143 s (given in Fig. 5.2b). First, the earthquake ruptured toward the northeast and then northwest of the epicenter. After a delay of the time, the rupture started from the epicenter again and extended toward the southwest with a distinct variation of the rupture direction at the latitude of 37°. **b** The relationship between released energy and rupture time. The dashed blue line indicates the ending time of the earthquake (143 s). There are three energy-radiation peaks at 23, 77, and 111 s. **c** The rupture time–distance relationship during the rupture process. First, the earthquake ruptured toward the northeast and then northwest of the epicenter at a slow rate of 0.92–1.10 km/s. The rupture front did not appear on the southwestern branch until 74 s. Then, the earthquake ruptured toward the southwest for 150 km at an average rate of 2.78–4.70 km/s from about 74–100 s. Finally, the earthquake ruptured to the southeast for a length of about 110 km on the third fault segment at an average rate of 1.51–1.62 km/s

time–distance relationship (Fig. 5.3c) shows that there were high-frequency energy-radiation peaks located at 14 km northeast, 34 km southwest, and 191 km southwest of the epicenter, corresponding to rupture times of 23, 77, and 111 s, respectively. Moreover, the time–distance relationship shows that in the first, 74 s the earthquake ruptured northeastwards over a length of 92 km at an average rupture speed of 0.92–1.10 km/s, while no rupture fronts appeared on the southwest branch. Due to non-uniquely determining the time corresponding to the main contributing wave energy within the 30-s window length, there is an uncertainty in the estimation of the rupture velocity (Kruger and Ohrnberger 2005). The first rupture front on the southwest branch started at 74.6 s at a distance of $\sim$28 km away from the epicenter. The earthquake continued to rupture southwestwards over

150 km at an average speed of 2.78–4.70 km/s until 100 s, when the rupture velocity suddenly decreased to 1.51–1.62 km/s. The rupture continued for an additional 110 km at this lower velocity and finally ended at 143 s. The rupture features described above imply that the Tohoku earthquake consisted of three rupture segments. Because the time lag between the first and second segments was about 74 s, we propose that the second segment might have been triggered by the first stage due to the localized increasing tectonic stress focused in the vicinity of the hypocenter, as a result of the rupture of the first segment.

In addition, there were three other prominent phenomena observed during the rupture process. First, the rupture front propagated at a very slow velocity along the strike direction during the first 25 s (Fig. 5.3c) with a very large energy-radiation peak northeast of the epicenter at 23 s (Fig. 5.3b). Second, the rupture direction of the first stage turned from northeast to northwest near the latitude of 38.5° at about 23 s (Fig. 5.3a). Third, there was a spatial gap of about 20 km during no more than 2 s and a distinct directional variation near the latitude of 37.0°, which separated the second segment from the third segment.

5.5 Discussion

The rupture evolution of an earthquake can be affected by the fault-zone properties (Di Stefano et al. 2011). In this study, most of the rupture features observed from the back-projection imaging (Fig. 5.2a) can also be associated with the lateral heterogeneities of the *P* and *S* wave velocity and Poisson's ratio (PR) structures which have been well investigated by tomographic imaging (Zhao et al. 1992, 2007) on the slab boundary. As shown in Fig. 5.4, there are two kinds of wave velocity and PR anomalies in the rupture zone. One is characterized by high *P* and *S* velocity and low-PR anomaly, which implies a more rigid medium. The other has a low *P* and *S* velocity and high-PR anomaly, which implies a more ductile medium. The hypocenter occurred in a region at a latitude of 38.32°, which is near the boundaries of high and low velocity and PR. The rupture extended northeastwards after the initiating earthquake. An energy peak was radiated when the rupture went through the boundary of velocity and PR anomalies into a more ductile region near 85.5° at 23 s. Due to the northwest-elongated anomalies distribution of the region, the rupture of the northern branch turned to the northwest direction. Next, the earthquake ruptured southwestwards. The second energy peak was released when the rupture went through a boundary of velocity and PR anomalies into a rigid region near the latitude of 38.0° at 77.0 s. Since the region of 37.0°–38.0° is more rigid, the rupture extended at a high rate across this region. Then, the rupture arrived at an anomaly boundary near the latitude of 37.0° at 110 s, when the energy peak of the third stage was radiated. Since the rupture zone of the third stage with a latitude below 37° has a southeast elongation of low *P* and *S* wave and high PR anomalies, the rupture direction turned to the southeast direction at a low rupture velocity. Therefore, due to the lateral heterogeneities of the *P* and *S* velocity and PR

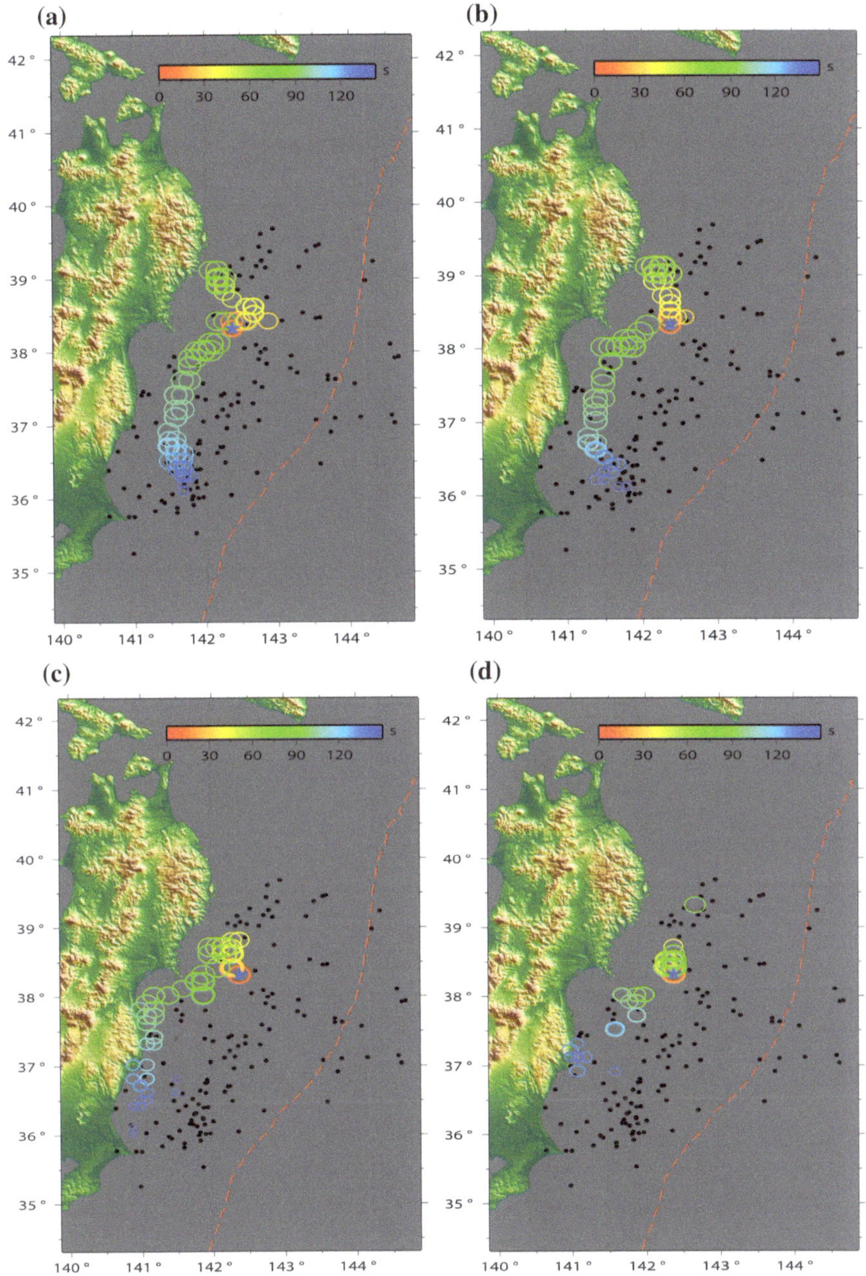

Fig. 5.4 Rupture imaging using high-pass filtered data with corner frequencies **a** 0.2 Hz, **b** 0.25 Hz, **c** 0.5 Hz, and **d** 1 Hz. As the corner frequency gets higher, the rupture fronts become closer to the northeast coast of Honshu, Japan

on the slab boundary of the northeast Japan arc, the rupture process of the 2011 Tohoku earthquake consisted of three subevents having different rupture velocities and energy-radiation peaks.

Not long after the earthquake, many source inversions of other types of data, such as tsunami, and broadband teleseismic data were carried out to investigate the slip of the earthquake (Lay et al. 2011). However, the areas of the large slip in these source models are quite different from most of the back-projection results (Koper et al. 2011). We think the main reason for the apparent inconsistency is that the main frequency of data used by back-projection is much higher than other types of

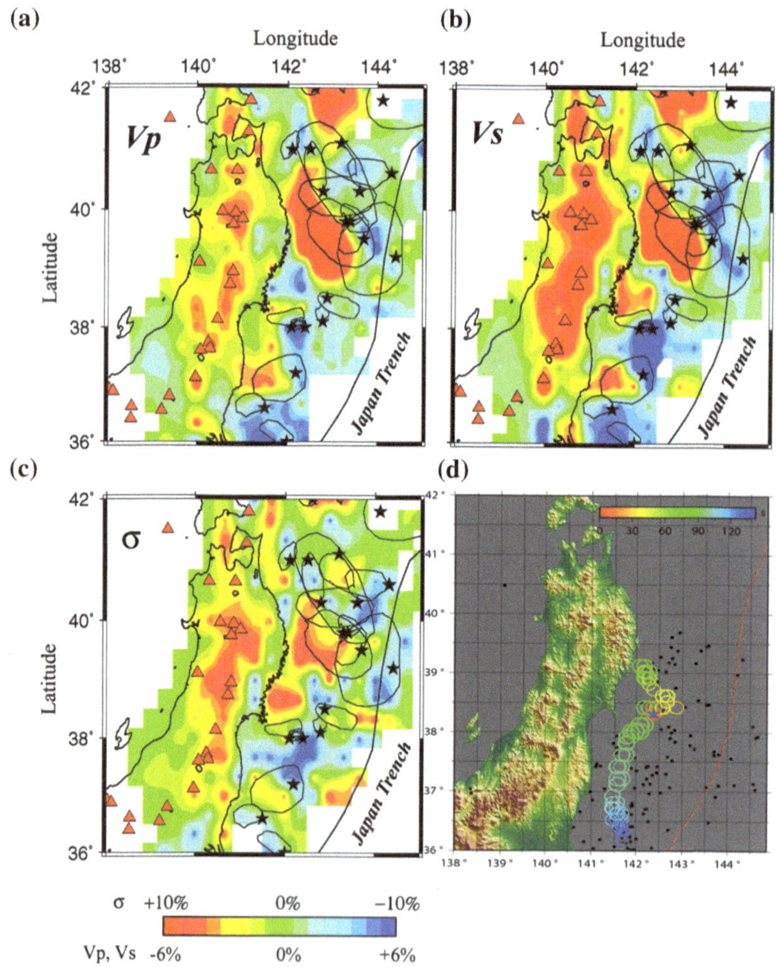

Fig. 5.5 Map view of **a** *P*- and **b** *S*-wave velocity and **c** Poisson's ratio images along the subduction interface and **d** The rupture process of the Tohoku earthquake. Figures (a, b, and c) are after Zhao et al. (2007)

data adopted by source inversions. The rupture features of the 2011 Tohoku Earthquake derived from back-projection have come from small-scale rupture structures. In contrast, the slip models from inversions are optimal solutions of the entire rupture process, in which small-scale features are smoothed. In addition, the characteristic frequency of the radiated signal varies during the rupture for a large earthquake (Kiser and Ishii 2011). Thus, source models inverted from different types of data with different frequency bandwidths reflect the rupture features of different segments and scales of the 2011 Tohoku earthquake.

The frequency-dependent feature has been observed in the 2010 Mw 8.8 Chilean earthquake by Kiser and Ishii (2011). For the 2011 Tohoku earthquake, the similar rupture feature was also revealed by Koper et al. (2011), Wang and Mori (2011), and Yao et al. (2011). The frequency-dependent feature in subduction zone megathrust earthquakes can be explained by asperities whose sizes change with depth (Lay et al. 2012). To confirm this feature, we carried out the back-projection using the data high-frequency filtered with corner frequencies of 0.2, 0.25, 0.5, and 1 Hz, respectively as shown in Fig. 5.5. The locations of the southern subevents become closer to the shore as the frequency of the data increases. As demonstrated by the frequency-dependent test in Chap. 2, the relative back-projection method cannot result in this kind of imaging feature. This indicates the frequency-dependent rupture feature does belong to the 2011 Mw 9.0 Tohoku earthquake.

5.6 Conclusions

In this paper, a two-step back-projection method, an array technique based on a non-plane-wave assumption, was applied to image the rupture process of the 2011 Mw 9.0 Tohoku earthquake. The rupture imaging resulting from this study shows that the 2011 Tohoku earthquake has a total rupture length of over 340 km and duration of at least 143 s, and ruptures to the northeast first and then to the southwest. Three energy-radiation peaks occurred at 23, 77, and 110 s, which correspond to 14 km northeast, 28 km southwest, and 191 km southwest of the epicenter, respectively. According to analyses of the lateral heterogeneities of *P*- and *S*-wave velocities in the rupture zone, the 2011 Tohoku earthquake consists of three rupture segments. The first one ruptured northeastwards and then northwards over ~90 km during the first 74 s at a speed of 0.92–1.10 km/s. Next, the second one ruptured southwestwards from 74 to 100 s with a length of over 150 km at a relatively high speed of 2.78–4.70 km/s, and might have been triggered by the increasing tectonic stress focused in the vicinity of the hypocenter as a result of the rupture of the first subevent. Eventually, the third one ruptured about 110 km southeastwards at an average speed of 1.51–1.62 km/s until the rupture ended at 143 s.

Some interesting rupture details of our model are presented here. The first rupture segment ruptured around the epicenter during the first 25 s after the initiating earthquake. The second segment ruptured at a relatively high velocity on the megathrust fault. A sudden rupture velocity variation appeared between the second and third subevents on the southwest branch near the latitude of 37°. Most of the rupture details are closely related to lateral heterogeneities of the *P* and *S* velocity and PR structures on the slab boundary of the northeast Japan arc. The imaged subevents are located deeper as the frequency of the data used in the back-projection increases. Based on the aforementioned observation, we can say that the rupture imaging infers that the 2011 Tohoku earthquake comprised three rupture segments (the first one on the northeast and the last two on the southwest).

References

Di Stefano, R., C. Chiarabba, L. Chiaraluce, M. Cocco, P. De Gori, D. Piccinini, and L. Valoroso. 2011. Fault zone properties affecting the rupture evolution of the 2009 (M 6.1) L'Aquila earthquake (central Italy): Insights from seismic tomography. *Geophysical Research Letters* 38 (10), L10310. doi:10.1029/2011GL047365.

Ishii, M., P.M. Shearer, H. Houston, and J.E. Vidale. 2005. Extent, duration and speed of the 2004 Sumatra-Andaman earthquake imaged by the Hi-Net array. *Nature* 435 (7044): 933–936.

Ishii, M., P.M. Shearer, H. Houston, and J.E. Vidale. 2007. Teleseismic P wave imaging of the 26 December 2004 Sumatra-Andaman and 28 March 2005 Sumatra earthquake ruptures using the Hi-net array. *Journal of Geophysical Research: Solid Earth (1978–2012)* 112: B11307.

Kennett, B.L.N., E.R. Engdahl, and R. Buland. 1995. Constraints on seismic velocities in the Earth from traveltimes. *Geophysical Journal International* 122 (1): 108–124.

Kiser E., and M. Ishii. 2011. The 2010 Mw 8.8 Chile earthquake: Triggering on multiple segments and frequency-dependent rupture behavior. *Geophysical Research Letters* 38 (7).

Koper, K.D., A.R. Hutko, and T. Lay. 2011. Along-dip variation of teleseismic short-period radiation from the 11 March 2011 Tohoku earthquake (Mw 9.0). *Geophysical Research Letters* 38: L21309.

Kruger, F., and M. Ohrnberger. 2005. Tracking the rupture of the Mw = 9.3 Sumatra earthquake over 1150 km at teleseismic distance. *Nature* 435 (7044): 937–939.

Lay, T., C.J. Ammon, H. Kanamori, L. Xue, and M.J. Kim. 2011. Possible large near-trench slip during the 2011 Mw 9. 0 off the Pacific coast of Tohoku Earthquake. *Earth, Planets and Space* 63 (7): 687–692.

Lay T., H. Kanamori, C.J. Ammon, et al. 2012. Depth-varying rupture properties of subduction zone megathrust faults. *Journal of Geophysical Research: Solid Earth (1978–2012)* 117: B04311.

Rost, S., and C. Thomas. 2002. Array seismology: Methods and applications. *Reviews of Geophysics* 40 (3): 1008.

Walker, K.T., M. Ishii, and P.M. Shearer. 2005. Rupture details of the 28 March 2005 Sumatra Mw 8.6 earthquake imaged with teleseismic P waves. *Geophysical Research Letters* 32: L24303.

Wang, D., and J. Mori. 2011. Frequency-dependent energy radiation and fault coupling for the 2010 Mw 8.8 Maule, Chile, and 2011 Mw 9.0 Tohoku, Japan, earthquakes. *Geophysical Research Letters* 38: L22308.

Xu Y., K.D. Koper, O. Sufri, et al. 2009. Rupture imaging of the Mw 7.9 12 May 2008 Wenchuan earthquake from back projection of teleseismic P waves. *Geochemistry, Geophysics, Geosystems* 10 (4): 1–17.

Yao, H., P. Gerstoft, P.M. Shearer, and C. Mecklenbrauker. 2011. Compressive sensing of the Tohoku-Oki Mw 9.0 earthquake: Frequency-dependent rupture modes. *Geophysical Research Letters* 38: L20310.

Zhang, H., and Z. Ge. 2010. Tracking the rupture of the 2008 Wenchuan earthquake by using the relative back-projection method. *Bulletin of the Seismological Society of America* 100 (5B): 2551–2560.

Zhao D., A. Hasegawa, and S. Horiuchi. 1992. Tomographic imaging of P and S wave velocity structure beneath northeastern Japan. *Journal of Geophysical Research: Solid Earth (1978–2012)* 97 (B13): 19909–19928.

Zhao, D., Z. Wang, N. Umino, and A. Hasegawa. 2007. Tomographic imaging outside a seismic network: Application to the northeast Japan arc. *Bulletin of the Seismo-logical Society of America* 97 (4): 1121–1132.

Chapter 6
Multi-fault Rupture and Dynamic Triggering During the 2012 Mw 8.6 Sumatra Earthquake

6.1 Introduction

On April 11, 2012 (08:38:37 UTC), a Mw 8.6 earthquake, which is the greatest strike-slip earthquake ever instrumentally recorded, occurred off the west coast of Sumatra Island. Nearly two hours later, a Mw 8.2 aftershock with a similar focal mechanism initiated approximately 180 km to SSW of the epicenter of the main-shock (2.311° N, 93.063° E, given by the US Geological Survey (USGS)), whereas a Mw 7.2 foreshock struck to the northeast near the epicenter on January 10, 2012 (Fig. 6.1). The 2012 Mw 8.6 Sumatra offshore earthquake occurred within the oceanic lithosphere in the north portion of the Wharton Basin, a diffuse boundary between the India and Australian plates surrounded by the Ninetyeast Ridge (90ER) to the west and the Sumatra trench to the northeast (Wiens et al. 1985). Deplus et al. (1998) evidently found that the deformation of the Wharton Basin is characterized by the reactivated N-S trending sinistral strike-slip fossil transform faults under the principal compressional regime-oriented NW-SE. Apart from the paleotransform faults, the oceanic lithosphere also deforms along the E-W striking abyssal hill fabric originating from the midocean spreading axes (Deplus 2001). The June 18, 2000, Wharton Basin Mw 7.8 earthquake in the south of the Wharton Basin has been determined to rupture along two conjugate planes (Robinson et al. 2001; Abercrombie et al. 2003), consistent with the aforementioned preexisting N-S paleotransform faults and the E-W abyssal hill fabric. Thus, it is meaningful to investigate what roles the preexisting weak structures played and how they interacted with each other during the 2012 Mw 8.6 earthquake.

The preliminary rupture pattern of the earthquake has been investigated by the finite fault model (Yue et al. 2012) and the back-projection method (Meng et al. 2012; Ishii et al. 2013). Here, we image the rupture process of the 2012 Mw 8.6 Sumatra offshore earthquake by back-projecting high-frequency teleseismic *P* waves (Ishii et al. 2005; Zhang and Ge 2010) from the European (EU), Australian (AU), and Japan (F-net) broadband seismic networks. To verify the reliability of the

H. Zhang, *Imaging the Rupture Processes of Earthquakes Using the Relative Back-Projection Method*, Springer Theses,
https://doi.org/10.1007/978-3-662-55239-1_6

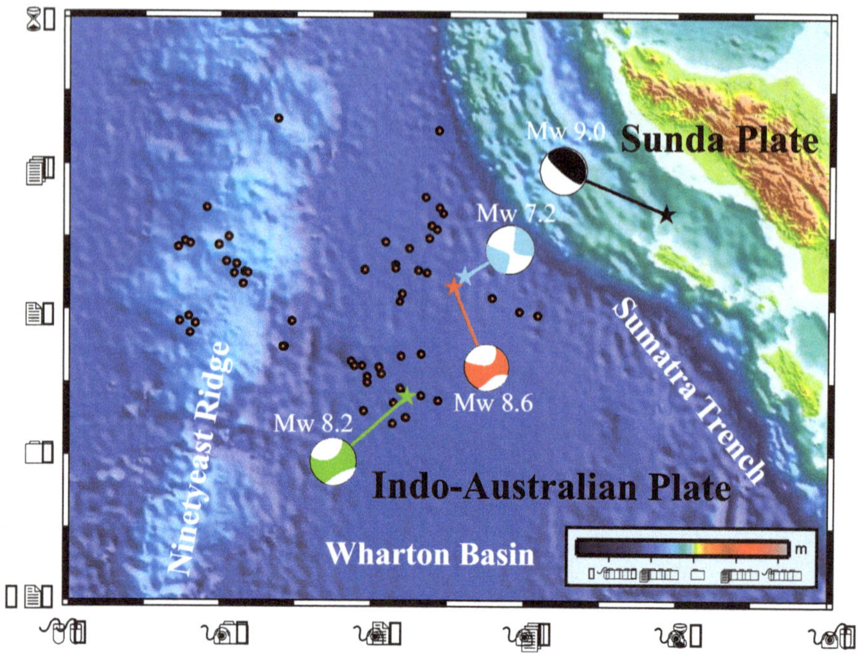

Fig. 6.1 Distribution of aftershocks and focal mechanisms of the December 24, 2004, Mw 9.0, January 10, 2012, Mw 7.2, April 11, 2012, Mw 8.6 and Mw 8.2 earthquakes given by gCMT. The red dots depict aftershocks with magnitudes larger than 4.5 on the first day. The stars indicate epicenters of the aforementioned four earthquakes given by USGS

results, an imaging test is applied for each network using synthetic data derived from the reflectivity method (Wang 1999). Finally, we propose a three-fault rupture model composed of two subparallel WNW-ESE trending faults and one NNE-SSW striking fault in between for the earthquake and investigate the triggering relationships among causative faults by calculating static Coulomb stress changes after the successive failure of the first two faults to try to understand the model that the 2012 Mw 8.6 earthquake ruptured three separated and conjugate faults.

6.2 Data and Methodology

To image the rupture of the 2012 Mw 8.6 offshore Sumatra earthquake, teleseismic *P* wave data from the EU, AU, and F-net seismic networks are analyzed using the relative back-projection method (Zhang and Ge 2010). These seismic networks have distinct azimuths (Fig. 6.2). Comprehensive analyses of the results from three seismic networks help to resolve the rupture process completely. Considering their signal-to-noise ratios (SNR) and coherences, vertical-component broadband

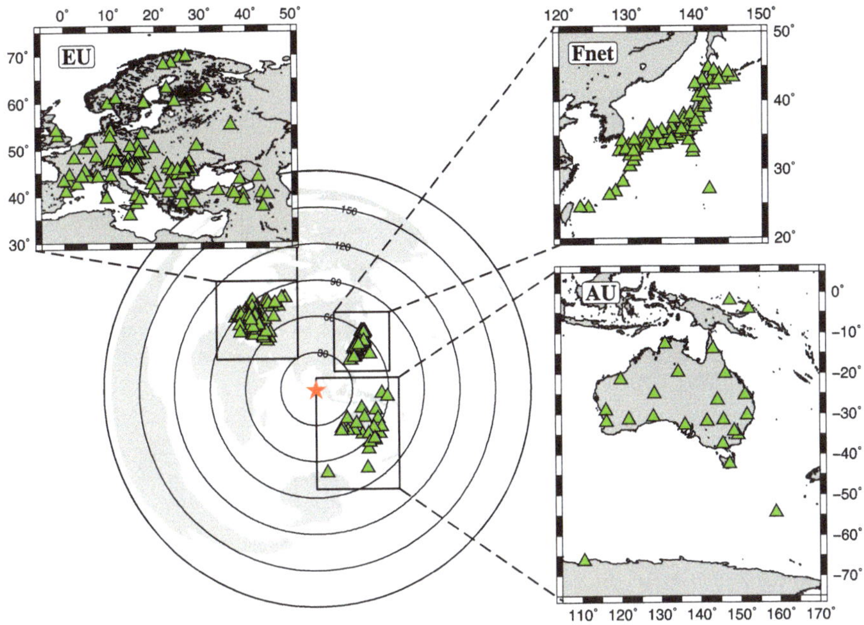

Fig. 6.2 Station distribution of the regional seismic networks EU, F-net, and AU. The green triangles indicate the stations, and the red star represents the epicenter of the 2012 Sumatra offshore earthquake

teleseismic *P* wave seismograms recorded at 113 stations of the EU, 24 stations of the AU, and 72 stations of the F-net (Fig. 6.3) are utilized in this study.

The back-projecting procedure is presented as follows. First, the *P* wave data are band-pass filtered in a high-frequency range of 0.2–2 Hz. The source region centered at the epicenter at a fixed focal depth of 22.9 km (given by USGS). Central points of the blocks are treated as potential rupture fronts. Calibrations to the deviations of the theoretical travel times, calculated using 1-D Earth velocity model AK135 (Kennett et al. 1995), from the observed arrival times are carried out by aligning the first 15 s of the *P* waveforms, using cross-correlation for all the stations with respect to the epicenter. Since the scale of the source region is much smaller than epidistances of the stations used in this study, travel time calibrations to the epicenter are applied to all the potential rupture fronts. Beams corresponding to the potential rupture fronts are formed by the 4th root stacking of the normalized seismograms. The powers of the potential rupture fronts are obtained by summing up the square of the corresponding beams in a 15-s-long sliding time window, shifted with a time step of 1 s, and then divided by the time period of 15 s. The potential rupture front with the maximum power at each time step is selected and assumed real, and then, its rupture time is obtained according to its location. Thus, the spatial and temporal evolution of the 2012 Mw 8.6 Sumatra offshore earthquake is derived from the EU, AU, and F-net and illustrated in Fig. 6.1.

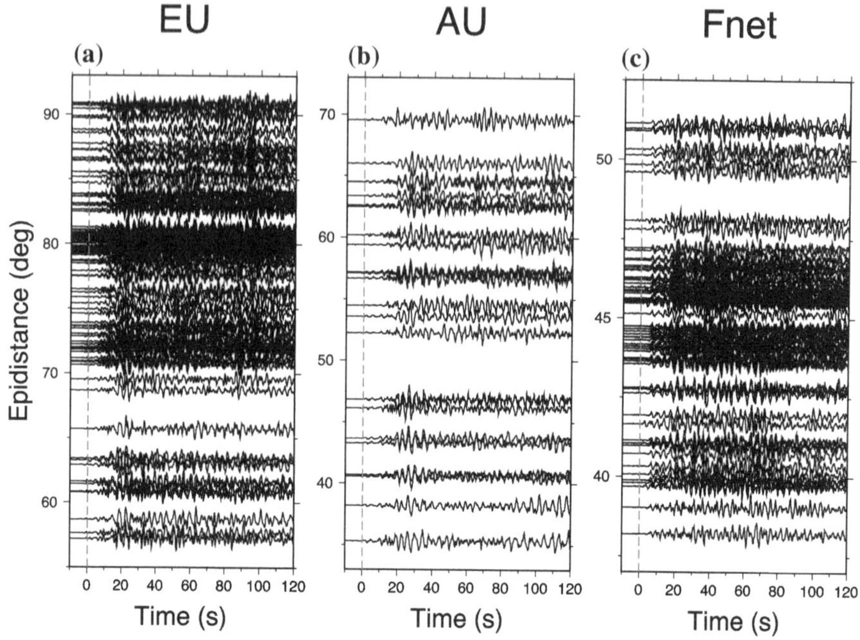

Fig. 6.3 Vertical-component seismograms recorded at these three networks, which are aligned (red dashed line) by cross-correlating the first 15-s-long waveforms

6.3 Results

The imaging results from network EU (Fig. 6.4a) demonstrate that the rupture of the earthquake was separated into three segments for a duration of at least 117 s. At the first segment, the rupture propagated ESEward away from the epicenter after the initiation of the quake until 35 s (Fig. 6.4d). It jumped to 70 km west of the epicenter and then expanded toward NNE for 42 s. Finally, the rupture initiated 170 km SW away from the epicenter and then extended WNWward during the last 40 s. The spatial distribution of the rupture fronts was in agreement with that of aftershocks on the first day except for those to the west on the 90ER (Fig. 6.4a). In addition, there existed a power releasing peak for each segment at 16 s, 64 s, and 95 s, respectively (Fig. 6.4d).

The final results derived from the AU (Fig. 6.4b) clearly indicates that the first segment linearly extended ESEward over a length of 105 km in the first 42 s, whereas the second segment mainly ruptured SSWward. The third one commenced at 80 s near the epicenter of the Mw 8.2 aftershock and about 180 km south of the epicenter of the mainshock and then propagated to WNW until 120 s, during which a peak in power was radiated at 105 s.

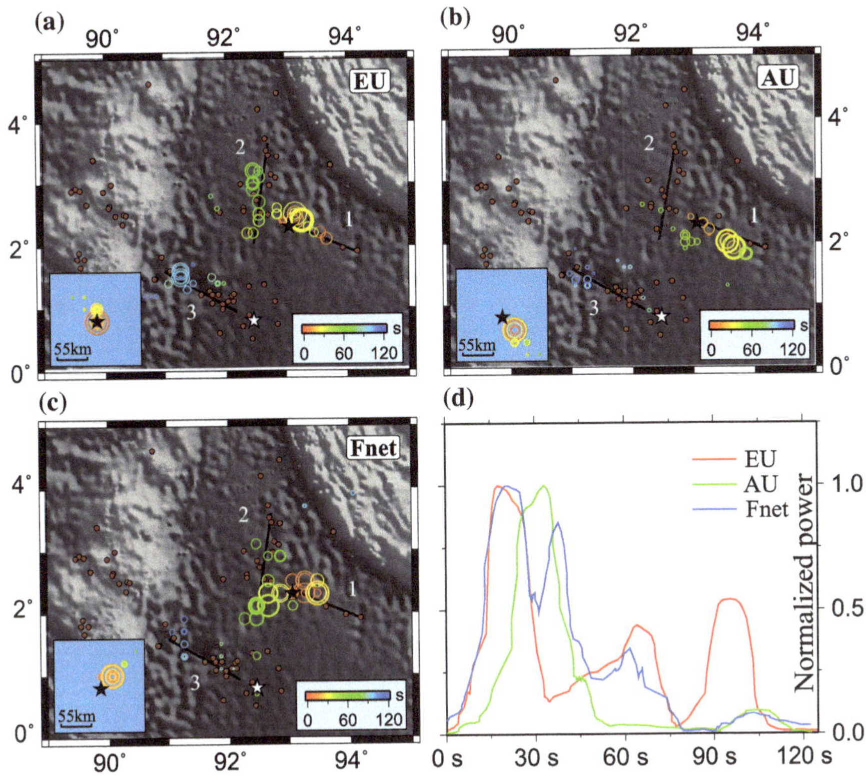

Fig. 6.4 Temporal and spatial evolution of the 2012 Mw 8.6 offshore Sumatra earthquake from **a** the EU, **b** the AU, and **c** the F-net. The circles indicate rupture fronts with colors representing rupture times and areas proportional to their power. The dark red dots denote aftershocks on the first day. The black lines labeled 1, 2, and 3 indicate the inferred traces of the first, second, and third faults, respectively. The black and white stars represent epicenters of the Mw 8.6 mainshock and Mw 8.2 aftershock. Insets with the same latitude and longitude scales illustrate the synthetic test for a point source at the epicenter (black star). **d** The normalized power of rupture fronts obtained by these three networks as a function of time

According to the results from the F-net, the first segment ruptured eastward for a duration of at least 30 s with a peak in power released at 16 s, whereas both the northern and southern branches of the second segment were imaged unambiguously in the rupture time range of 32–80 s and released power with maxima at 40 and 62 s, respectively (Fig. 6.4c). The southern branch initiated at 32 s and terminated at 52 s after the first rupture front of the northern branch showed up (Fig. 6.4d). This indicates that the second segment probably ruptured SSWward first and then NNEward. Finally, the third segment initiated near the epicenter of the Mw 8.2 aftershock and then extended WNWward in the last 40 s with a peak in power at 104 s.

To verify the reliability of the rupture pattern and interference from other phases such as *pP*, *sP*, and *PcP*, a test is carried out to back-project 120-s-long synthetic data, which are calculated at the AU, EU, and F-net (Fig. 6.5) using the reflectivity method (Wang 1999). The focal mechanism (strike 20, dip 64, rake 1) given by the global Centroid Moment Tensor (gCMT) and normalized square half-sinusoid wavelets with a duration of 1.2 s are utilized to calculate the synthetic seismograms in the one-dimensional Earth velocity model AK135 (Kennett et al. 1995). Synthetic data calculated from a point source located at the epicenter are back-projected to the source region. Most of the rupture fronts from the EU are located at the epicenter, whereas those from the AU and F-net are dragged toward orientations of the corresponding networks and have a mislocation of 31 km to the epicenter (insets of Fig. 6.4), which results from the amplitudes of depth phases such as *pP*, and *sP* recorded at the AU and F-net are larger than those of the direct *P* waves (Fig. 6.5). This indicates that the imaging results from the EU are more reliable than those from the AU and F-net. Moreover, the delay of the first peak from network AU (Fig. 6.4d) is caused by the interference of depth phases. However, since location offsets among rupture fronts are much larger than the mislocation from *pP* and *sP* phases, results derived from the AU and F-net still reflect a real rupture feature of the earthquake to a great extent.

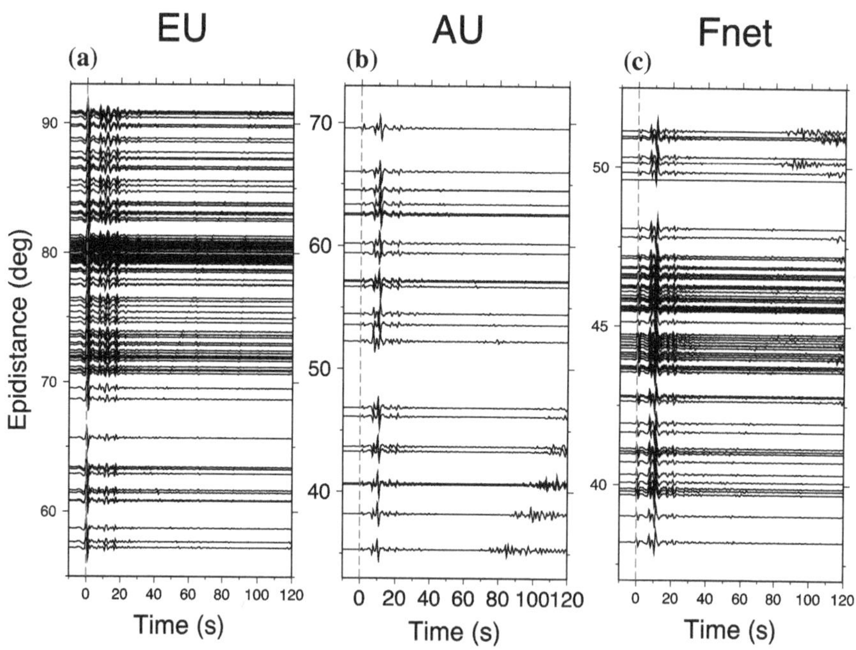

Fig. 6.5 Vertical-component synthetic seismograms received at these three networks. These seismograms are calculated from a point source at the epicenter using the reflectivity method (Wang 1999) and are aligned (red dashed line) by the first *P* arrivals

The magnitudes of high-frequency coherent signals play a significant role in the imaging process and account for the differences between the results derived from these three networks. They are greatly influenced by rupture directivity and radiation pattern associated with the azimuths of the networks. As the previous description to the rupture directionality of the earthquake shows, the rupture propagated ESEward on the first fault but WNWward on the third fault and extended SSWward first and then NNEward along the second fault. Signals radiated from the first fault and recorded at EU have lower frequencies and longer durations with respect to those received at F-net. Therefore, these signals overlap the beginning part of those with weaker amplitudes from the southern branch of the second fault for EU. This leads to a 3-s delay to the initiation of the southern branch of the second fault for network EU relative to F-net. The interference of depth phases (*pP* and *sP*) makes the power of rupture fronts of the southern branch with respect to those of the first fault from network F-net larger than those from EU due to the effect of radiation pattern as shown in insets of Fig. 6.4a, c. In addition, the rupture directivity of the northern branch causes the differences between the images derived from networks EU and AU. According to the preceding analysis of sensitivity to the rupture, the initiation time of the northern branch is identified to be 47 s from network EU, while the ending time of the southern branch is determined to be 52 s from network F-net.

The major patterns of *P* wave back-projected rupture images in terms of the earthquake obtained by Meng et al. (2012), using data from the European and Japanese (Hi-net) seismic networks, and by Satriano et al. (2012), employing data from the European network, are similar to those derived from the EU and F-net in this study, except for some details and their fourth segment from 120 to 160 s, which is unresolved in our images. The imaging difference results from different band-passing frequency ranges applied to data and distinct strategies to mitigate the imaging artifacts. Two parallel WNW-ESE striking causative faults are resolved from long-period seismic data by Duputel et al. (2012) adopting multiple point source inversion, which is consistent with the rupture images in this study. However, preliminary finite fault models (Hayes 2012; Shao et al. 2012) present a very compact slip distribution around the epicenter with a rupture length of no more than 150 km on a given causative planar fault. The one-fault assumption is responsible for the differences between the finite fault model and the back-projected imaging.

Combining analyses and tests to the rupture images from networks EU, AU and F-net, comparison of the results with those from other groups and approaches, and spatial distribution of aftershocks and preexisting weak structures, we propose that the 2012 Mw 8.6 Sumatra offshore earthquake ruptured three separated and conjugate faults, two subparallel WNW-ESE trending dextral faults, and one NNE-SSW striking sinistral fault in between (Fig. 6.6), with specific parameters as listed in Table 6.1.

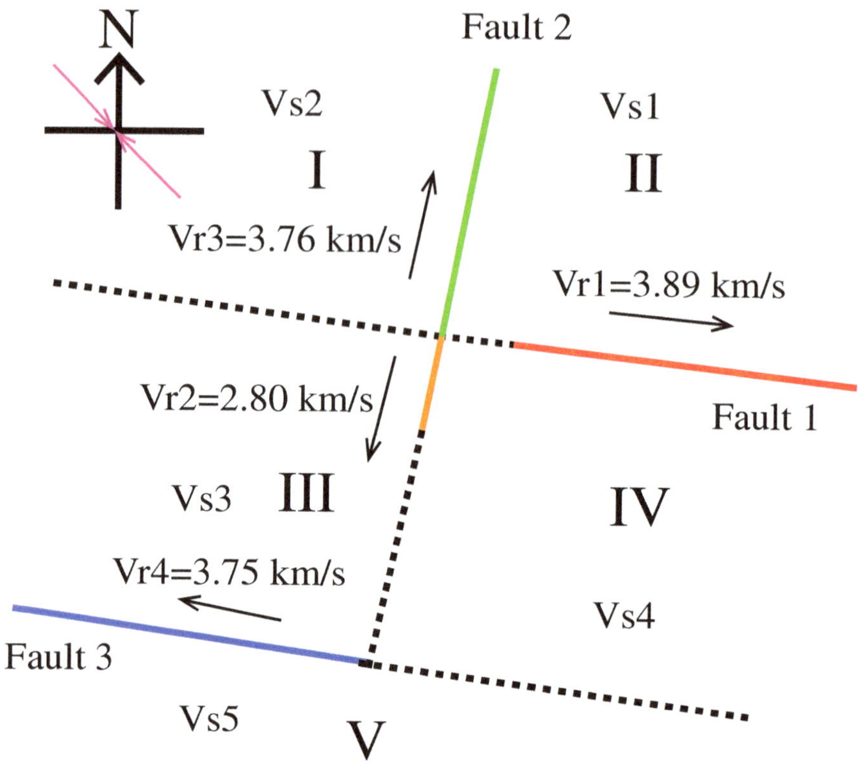

Fig. 6.6 Schematic rupture model of the 2012 Sumatra earthquake. The red and blue lines indicate fault 1 and 3, respectively, while the orange and green lines depict the southern and northern branches of fault 2, respectively. The dashed lines are the extension of the faults without rupture. The purple arrow pair denotes the direction of the compressive tectonic stress. The black arrows represent the rupture directions of the faults. Vr1–4 are the rupture velocities of fault 1, the southern and northern branches of fault 2, and fault 3, respectively. Vs1–5 represent the shear wave velocities of source region I–V. If the shear velocities of these regions satisfy the expression of Vs1 >= Vs5 >= Vs4 >= Vs2 > Vs3, the rupture velocities could be those illustrated in this figure, according to the relationship between shear wave and rupture velocities experimentally implied by Xia et al. (2005)

Table 6.1 Source Parameters and Rupture Model of the 2012 Sumatra Offshore Earthquake

Fault #	Seismic moment (%)	Strike (deg)	Rupture direction	Length (km)	Rupture time (s)	Duration (s)	Rupture velocity (km/s)
1	49.67	112	ESE	136	0–35	35	3.89
2S	12.17	11	SSW	56	32–52	20	2.80
2N	26.96	11	NNE	124	47–80	33	3.76
3	11.20	119	WNW	150	80–120	40	3.75

6.4 Static Triggering Among Causative Faults

The aftershocks following an earthquake are usually triggered by static Coulomb failure stress changes (Freed 2005). Here, the static Coulomb failure stress changes in a homogeneous isotropic elastic half-space for a finite rectangular source (Okada 1992) are calculated using the software Coulomb 3.3 (Lin and Stein 2004; Toda et al. 2005). The detailed procedure of the calculation is given below.

To accomplish the analysis, we employ geometries and focal mechanisms of three causative faults as listed in Table 6.1. The bottom of the fault plane is set to 30 km owing to the fact that frictional failure in oceanic lithosphere could occur below 600 °C implied by the laboratory studies of olivine (Boettcher et al. 2007). Dip angles of these three faults are set to 90° in our model for simplification, since those of focal mechanisms of the earthquake given by gCMT and USGS are in a range of 64°–87°. The total seismic moment of 8.96 × 10^{21} Nm given by gCMT is adopted and then partitioned for three causative faults (Table 6.1), according to

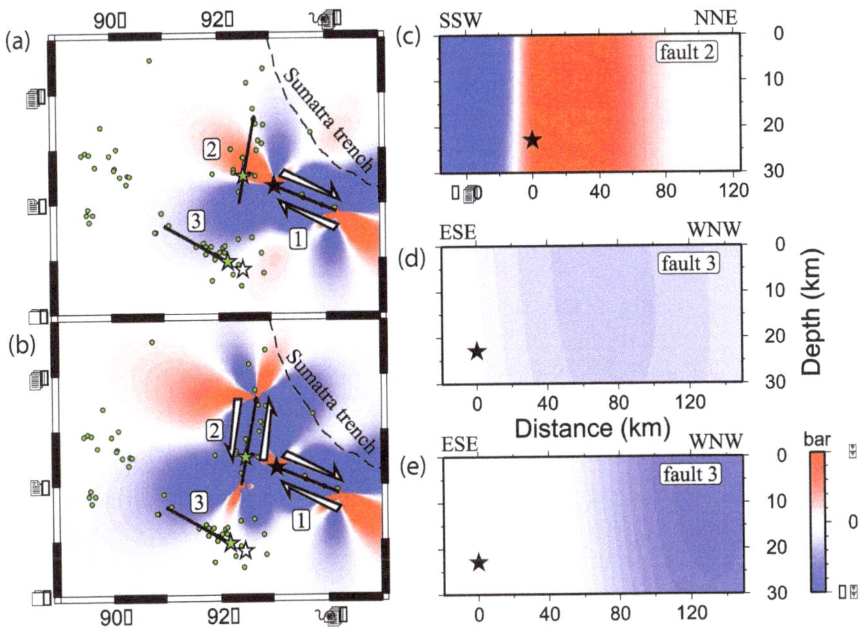

Fig. 6.7 Static Coulomb stress changes at optimal orientations after the failures of **a** the first fault and then **b** the second fault. The projection of the static stress changes on **c** the second fault and **d** the third fault after the failure of the first fault and on **e** the third fault after the failure of the second fault. The color bar indicates the static Coulomb stress change. The three black lines labeled 1, 2, and 3 denote surface traces of three sequential conjugate causative faults, respectively. The black and white stars represent epicenters of the Mw 8.6 mainshock and the Mw 8.2 aftershock, respectively. The green stars depict the initial failure points of the second and third faults. The green dots indicate aftershocks. The black stars in Fig. 6.2c–e indicate the initial points of the failure on the corresponding faults

their corresponding energy integrating normalized power with time (Fig. 6.4d). The static Coulomb failure stress changes are calculated in the form $\Delta CFS = \Delta\tau + \mu'\Delta\sigma_n$ (King et al. 1994; Stein 1999) under the assumption of uniform slip, where is $\Delta\tau$ the shear stress change, $\Delta\sigma_n$ is the normal stress change, and μ' is the effective friction coefficient. Since the range of the effective friction coefficient in the coseismic stress changes is 0.0–0.75 (Freed 2005), we choose a median value of 0.4. The Poisson's ratio and Young's modulus are set to 0.25 and 8×10^{10} Pa, respectively. The 2-D horizontal source region fixed at a depth of 22.9 km and vertical receiver faults are gridded into blocks of 5.56×5.56 km^2 and 1×1 km^2, respectively. Finally, we obtain the static Coulomb stress changes at optimal orientations and on the receiver faults in two cases (one is after the failure of the first fault, and the other is after the second fault) as illustrated in Fig. 6.7.

After the failure of the first dextral fault, the static Coulomb stress increases at the initial points of the failure on the second and third faults are 2.89 and 0.66 bars, respectively (Fig. 6.7c, d). The failures of the first and second faults lead to the static Coulomb stress increase of 0.26 bars at the initial point of the failure on the third fault (Fig. 6.7e). This means that the rupture of the second fault is probably triggered by the prominent increase in the static Coulomb failure stress stemming from the failure of the first fault. Also, the rupture of the second fault leads to a static Coulomb stress increase greater than the triggering threshold of 0.1 bars (Stein 1999), which possibly promotes the third right-lateral fault to fail.

6.5 Discussion and Conclusions

The triple-fault rupture model from the straightforward back-projection method is consistent with preexisting weak structures (Deplus et al. 1998), complicated aftershock distribution (Fig. 6.4), and the source model derived from multiple point source inversion (Duputel et al. 2012). The three conjugate faults are separated from each other; the initial failure point of the second fault is 68 km WNW away from the western end of the first fault, whereas that of the third fault is located 120 km SSW from the southern end of the second fault near the epicenter of the Mw 8.2 aftershock. Rupturing multiple faults has been discovered during other earthquakes. For instance, in the same tectonic context as the mainshock, the June 18, 2000, Wharton Basin Mw 7.8 earthquake was observed to rupture along two conjugate faults (Robinson et al. 2001; Abercrombie et al. 2003).

The average rupture velocities of faults are about 3.75–3.90 km/s except for the southern branch of the second fault (2.80 km/s) as listed in Table 6.1 and illustrated in Figure S5. The appropriate shear wave velocity in the source region should be 3.90–4.49 km/s for a focal depth of 22.9 km (given by USGS) within the oceanic lithosphere, which is extracted from a 1-D shear wave velocity model combining Crust 2.0 (Bassin et al. 2000) with AK135 (Kennett et al. 1995). Moreover, it is common for strike-slip faults with steep dip angles to be ruptured at a high subshear wave velocity. In addition, the lower rupture velocity of southern branch of the second fault could be explained by the lower shear velocity of the region to the west of the fault according to the laboratory studies of rupture directionality (Xia et al. 2005) that rupture velocity might positively correlate with shear wave velocities of media around the fault. The lower shear velocity indicates that the region may have a high-heat anomaly and be more rheological than other regions. Thus given, the fault of the southern branch would be weaker than the northern branch. This explains the observation that after the static triggering of the second fault from the first fault (Fig. 6.7d), the rupture firstly extended on the southern branch, whose static Coulomb stresses decreased though. The rupture then jumped on the northern branch of the second fault because of the failure of its SSWward end caused by the stress increase resulting from the 15-s-long faulting of the southern branch.

Besides the static Coulomb stress changes, dynamic stress increases resulted from the passage of seismic waves have pronounced effects on triggering the failure of receiver faults. Dynamic stress increases at distances more than one fault length may be ten times larger than the static Coulomb stress changes (Stein 1999). In the direction of rupture propagation, dynamic stress increases at an order of magnitude larger than in other directions (Kilb et al. 2000). Taken the static Coulomb stress changes as the references, the dynamic stress changes could be roughly estimated to the first-order extent. For the 2012 Mw 8.6 Sumatra offshore earthquake, the dynamic stress increase at the initial failure point of the second fault induced by the unilateral ESEward rupture of the first fault (68 km ESE of the second fault) could be comparable to the static Coulomb stress change. This means the dynamic stress change probably plays the same significant role as the static in triggering the second fault. In contrast, the failure initiation of the third fault is located at the front of the rupture direction of the southern branch of the second fault about one fault length (120–175 km) away. Dynamic triggering should play a more significant role than the static one in this case. To assess the effect of the dynamic triggering from the southern branch of the second fault to the third fault, we synthetically calculate three-component acceleration seismograms (Fig. 6.8) at the initial point of the third fault with a depth of 22.9 km in a combined velocity model of Crust 2.0 (Bassin et al. 2000) and AK135 (Kennett et al. 1995). The rupture of the southern branch of the second fault is simulated by the SSWward proceeding of a vertical linear source with a depth range of 0–30 km at a uniform rupture velocity of 2.8 km/s. The amplitudes of transverse component are much larger than those of radial and vertical components, which excludes the triggering from P and Rayleigh waves. There are several acceleration peaks of S waves forcefully acting on the initial point of the third fault toward the WNW before its rupture initiation (80 s). In addition, the velocities of Love waves at shallow depths are almost equal to those of *S* waves. Thus, the third fault is primarily dynamically triggered by the passage of S or even Love waves radiated from the southern branch of the second fault, although the static Coulomb stress increase cannot be excluded.

The 2012 Mw 8.6 Sumatra offshore earthquake ruptured three sequential faults over a time period of at least 120 s, which is inferred from the high-frequency back-projected teleseismic *P* wave data. On the first fault, the rupture propagated

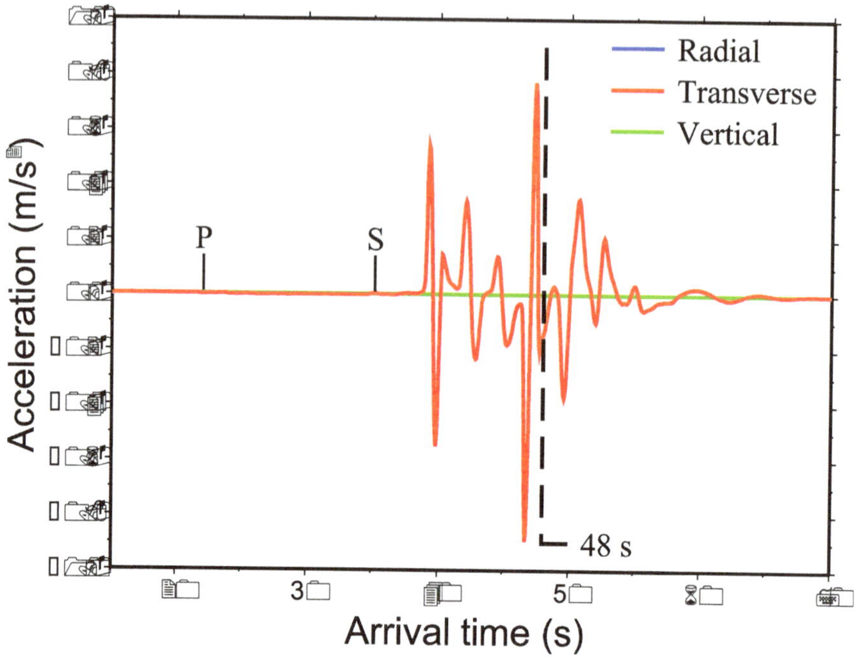

Fig. 6.8 Three-component synthetic acceleration seismograms radiated from the southern branch of the second fault and recorded at the initial point of the third fault in the RTZ coordinate system. The blue, red, and green lines indicate the radial, transverse, and vertical components, which have positive directions of SSW, ESE, and Up, respectively. The “P” and “S” are labeled where the first arrivals of *P* and *S* waves are located, respectively. The dashed line represents the initial time of the third fault, which is 80 and 48 s after those of the first and second faults, respectively

ESEward over a length of 136 km and a time duration of 35 s at an average velocity of 3.89 km/s after the onset of the quake. Then, the rupture initiated at 32 s and commenced at the intersected point of the first and second faults, 68 km WNW from the epicenter, and firstly propagated SSWward over a length of 56 km within a period of 20 and 15 s later NNEward over a length of 124 km for a duration of 33 s at average rates of 2.80 and 3.76 km/s, respectively. Finally, the rupture of the third fault commenced near the epicenter of the Mw 8.2 aftershock and then continued WNWward during the last 40 s at an average speed of 3.75 km/s. In addition, the failure of the second fault was triggered by the static Coulomb and dynamic stress increases from the failure of the first fault. The rupture of the third fault was probably dynamically triggered by S or even Love waves from the southern branch of the second fault, even though the static Coulomb stress increase from the failure of the second fault partly stimulated the failure initiation. Note that it is essential to use multiple networks in different orientations jointly in order to obtain the reliable and complete rupture pattern of events with highly azimuth-dependent signals such as the 2012 Sumatra offshore earthquake.

References

Abercrombie, R.E., M. Antolik, and G. Ekstrom. 2003. The June 2000 Mw 7.9 earthquakes south of Sumatra: Deformation in the India-Australia Plate. *Journal of Geophysical Research* 108 (B1): 2018.

Bassin, C., G. Laske, and G. Masters. 2000. The current limits of resolution for surface wave tomography in North America. *EOS Transactions, AGU* 81: F897.

Boettcher, M.S., G. Hirth, and B. Evans. 2007. Olivine friction at the base of oceanic seismogenic zones. *Journal of Geophysical Research: Solid Earth* 112 (B1).

Deplus, C. 2001. Indian Ocean actively deforms. *Science* 292 (5523): 1850–1851.

Deplus, C., M. Diament, H. Hebert, et al. 1998. Direct evidence of active deformation in the eastern Indian oceanic plate. *Geology* 26 (2): 131–134.

Duputel, Z., H. Kanamori, V.C. Tsai, L. Rivera, L. Meng, J.P. Ampuero, and J.M. Stock. 2012. The 2012 Sumatra great earthquake sequence. *Earth and Planetary Science Letters* 351–352: 247–257.

Freed, A.M. 2005. Earthquake triggering by static, dynamic, and postseismic stress trans- fer. *Annual Review of Earth and Planetary Sciences* 33: 335–367.

Hayes, G.P. 2012. Finite fault model: Preliminary result of the Apr 11, 2012 Mw 8.6 earthquake off west coast of the northern Sumatra, report, U.S. Geol. Surv., Reston, Va. Available at http://www.usgs.gov.

Ishii, M., E. Kiser, and E.L. Geist. 2013. Mw 8.6 Sumatran earthquake of 11 April 2012: Rare seaward expression of oblique subduction. *Geology* 41 (3): 319–322.

Ishii, M., P.M. Shearer, H. Houston, and J.E. Vidale. 2005. Extent, duration and speed of the 2004 Sumatra-Andaman earthquake imaged by the Hi-Net array. *Nature* 435 (7044): 933–936.

Kilb, D., J. Gomberg, and P. Bodin. 2000. Triggering of earthquake aftershocks by dynamic stresses. *Nature* 408 (6812): 570–574.

King, G.C.P., R.S. Stein, and J. Lin. 1994. Static stress changes and the triggering of earth- quakes. *Bulletin of the Seismological Society of America* 84 (3): 935–953.

Kennett, B.L.N., E.R. Engdahl, and R. Buland. 1995. Constraints on seismic velocities in the Earth from traveltimes. *Geophysical Journal International* 122 (1): 108–124.

Lin, J., and R.S. Stein. 2004. Stress triggering in thrust and subduction earthquakes and stress interaction between the southern San Andreas and nearby thrust and strike- slip faults. *Journal of Geophysical Research: Solid Earth (1978–2012)* 109: B02304.

Meng, L., J.-P. Ampuero, J. Stock, et al. 2012. Earthquake in a maze: compressional rupture branching during the 2012 Mw 8.6 Sumatra earthquake. *Science* 337 (6095): 724–726.

Okada, Y. 1992. Internal deformation due to shear and tensile faults in a half-space. *Bulletin of the Seismological Society of America* 82 (2): 1018–1040.

Robinson, D.P., C. Henry, S. Das, and J.H. Woodhouse. 2001. Simultaneous rupture along two conjugate planes of the Wharton Basin earthquake. *Science* 292 (5519): 1145–1148.

Satriano, C., E. Kiraly, P. Bernard, and J.-P. Vilotte. 2012. The 2012 Mw 8.6 Sumatra earthquake: Evidence of westward sequential seismic ruptures associated to the reactivation of a N-S ocean fabric. *Geophysical Research Letters* 39: L15302. https://doi.org/10.1029/2012GL052387.

Shao, G.F., X. Li, and C. Ji. 2012. Preliminary result of the Apr 11, 2011 Mw 8.64 Sumatra earthquake. Available at www.geol.ucsb.edu.

Stein, R.S. 1999. The role of stress transfer in earthquake occurrence. *Nature* 402 (6762): 605–609.

Toda, S., R.S. Stein, K. Richards-Dinger, and S.B. Bozkurt. 2005. Forecasting the evolution of seismicity in southern California: Animations built on earthquake stress transfer. *Journal of Geophysical Research* 110: B05S16.

Wang, R. 1999. A simple orthonormalization method for stable and efficient computation of Green's functions. *Bulletin of the Seismological Society of America* 89 (3): 733–741.

Wiens, D.A., C. DeMets, R.G. Gordon, et al. 1985. A diffuse plate boundary model for Indian Ocean tectonics. *Geophysical Research Letters* 12 (7): 429–432.

Xia, K., A.J. Rosakis, H. Kanamori, and J.R. Rice. 2005. Laboratory earthquakes along inhomogeneous faults: Directionality and supershear. *Science* 308 (5722): 681–684.

Yue, H., T. Lay, and K.D. Koper. 2012. En echelon and orthogonal fault ruptures of the 11 April 2012 great intraplate earthquakes. *Nature* 490 (7419): 245–249.

Zhang, H., and Z. Ge. 2010. Tracking the rupture of the 2008 Wenchuan earthquake by using the relative back-projection method. *Bulletin of the Seismological Society of America* 100 (5B): 2551–2560.

Chapter 7
Discussion, Conclusions, and Perspectives

7.1 Discussion

7.1.1 Method

The method used in this study is the relative back-projection method (RBPM), which is one of the generalized array techniques. This method was modified from the traditional back-projection method (BPM). The RBPM images the subevents in the time window of waveforms at the reference station, and the traditional BPM images the potential subevent with the maximum power as the real subevent in the rupture time of the earthquake. The imaging of the traditional BPM suffers from severe influence from the large high-frequency (HF energy bursts). Due to the trade-off between rupture time and location along the ray path, these HF energy bursts can cause an imaging maximum swimming along the ray path. Since the RBPM operates the data in the time line of the seismogram at the reference time, the HF energy bursts would not affect the imaging beyond the scopes of these HF energy bursts; this significantly mitigates the "swimming" artifacts in the traditional back-projection imaging and makes the imaging area closer to the actual location of the subevent. Because of the travel time of the reference station is used to do the time calibration, the travel time difference at the same station for the RBPM is less than that for the traditional BPM. This is the reason that the 80% contour of the maximum power of the relative back-projection imaging is larger than that of the traditional back-projection imaging. Therefore, the potential subevent with the maximum power was chosen as the real subevent for the RBPM.

Besides the back-projection imaging, the maximum likelihood method (Capon 1969), multiple signal classifying method (Schmidt 1986) and compressive sensing (Yao et al. 2011), used to determine the arrival azimuths of the impinging signals, were applied in researches on earthquake source recently. The principle of these approaches to find sources of the impinging signals is to set up a function to make it being the maximum or minimum in the impinging direction. In theory, multiple

H. Zhang, *Imaging the Rupture Processes of Earthquakes Using the Relative Back-Projection Method*, Springer Theses,
https://doi.org/10.1007/978-3-662-55239-1_7

signals at the same time can be recognized when they significantly differ from each other. For the earthquakes, the difference between the signals from different subevents is typically not large enough to achieve this. Moreover, these approaches may consume much more computational resources on solving eigenvectors or iteratively searching the subevents. This may be an issue for quick earthquake response. Thus, the application of the array techniques depends on the research goals.

7.1.2 Data Used

The data used in this book are teleseismic *P* waves. Besides the *P* waves, other phases such as *S* waves (Allmann and Shearer 2007) and even surface waves (Roten et al. 2012; Yue et al. 2012) can be used in the back-projection. According to the Rayleigh criterion, the minimum angle to discriminate two objects is expressed as

$$\theta_{\min} = 1.22\frac{\lambda}{D},$$

where λ is the wavelength, and D is the diameter of the lens in Optics and is the aperture of the seismic array in this study. Based on this equation, the spatial resolution of an array is inversely proportional to the wavelength. For the given array, the resolution of the *P* wave, *S* wave, and surface wave sequentially decreases. However, the amplitude of the three waves increases sequentially. Due to attenuation and scattering, the frequency of the three waves decreases sequentially. Moreover, the surface waves have much lower frequency than the two body waves (Yue et al. 2012). Since *S* waves arrive later than *P* waves, picking up *S* arrivals is much more difficult than picking up *P* arrivals; this increases the errors in the initial time calibration for the *S* waves and results in more uncertainties in imaging the subevents. Since different kinds of waves have different perspectives on the earthquake sources, the combination of these waves could provide comprehensive constraints on the rupture of the earthquakes. In the application of the back-projection imaging to the rupture of earthquakes, the data used to depend on the number of stations, frequency band, signal-to-noise ratio, and coherence of the waveforms.

7.1.3 Frequency-Dependent Rupture

The frequency-dependent rupture describes a rupture feature that the frequency content of the waveforms increases with the focal depth. This rupture feature has been observed in subduction zone megathrust earthquakes like the 2010 Chilean earthquake and the 2011 Tohoku earthquake (Kiser and Ishii 2011; Lay et al. 2012; Yao et al. 2013).

For a large earthquake, the high-frequency back-projection imaging rupture model is typically different from the slip model derived by inverting longer period body waves or surface waves (Lay et al. 2012). High-frequency seismic energy is released when the rupture velocity changes like acceleration, deceleration, commencing and stopping of the rupture (Madariaga 1983, 1977), and healing of fault (McLaskey et al. 2012). The changing behaviors of the rupture velocity are associated with the rupture front. Since the fault healing is relatively rare, the high-frequency seismic energy released by the earthquake mainly stems from the rupture front.

There are two conceptual models to describe the rupture of earthquakes. The first one is the barrier model proposed by Das and Aki (1977). When the rupture front arrives at a barrier, the rupture is usually stopped. The stopping phase could release a large amount of high-frequency seismic energy. Thus, the high-frequency sources may mark the edges of the barriers on the fault. The subfaults releasing large amount of seismic moment represent the region storing much strain energy. Thus, the locations of the high-frequency sources and low-frequency sources are different from each other.

The second model is the asperity model (Lay and Kanamori 1981; Rudnicki and Kanamori 1981). In the asperity model, the fault strength is distributed evenly on the fault, which is different from the barrier model, in which the stress is evenly distributed on the fault (Aki 1984). The asperities are the high-stress regions on the fault. The frequency of the waves released during the earthquake is closely associated with the stress on the asperity as well as the asperity size. When the rupture arrives at an asperity, the initial and ending failure of the asperity would release a lot of high-frequency seismic energy. The continuous failure of the asperity releases a large amount of low-frequency seismic energy. Thus, the depth difference between HF- and LF- sources can also be explained by the asperity model.

For the frequency-dependent rupture feature, Lay et al. (2012) proposed an asperity model combined with the rate and state relationship (Dieterich 1979; Scholz 1998). Specifically, the asperity size is related to the status of temperature and pressure. When the focal depth increases, the temperature and pressure accordingly increase; this leads to the smaller asperity size, when close to the bottom of the seismogenic zone. The rupture of the smaller asperity releases seismic energy dominated by higher frequency content. This indicates the higher frequency sources are located deeper on the fault plane.

7.2 Conclusions

Beginning with the representation theorem, the principle of the generalized array techniques was proofed in a mathematically way. After that, five widely used array techniques were introduced. The relative back-projection method used in this book was modified from the traditional back-projection method and elaborated in detail to separately resolve the locations and rupture times of the subevents during the earthquakes. To demonstrate the effective scope of the method, series of synthetic tests have been conducted using an N-S striking linear array. The spatial resolutions along the range and azimuth directions have been explored accordingly. The tests show that the azimuthal resolution is higher than the range resolution. Moreover, the relative back-projection imaging area is closer to the real location of a subevent relative to the traditional back-projection imaging results. More importantly, there are no "swimming" artifacts in the relative back-projection imaging. Thus, the relative back-projection can accurately image the rupture of the earthquakes without suffering the "swimming" artifacts like the traditional back-projection.

To demonstrate its effectiveness, the relative back-projection method was first applied to image the rupture process of the 2008 Wenchuan earthquake using the teleseismic *P* waves. Because the realistic data are typically more complex than the synthesized data, there exist other kinds of artifacts. To remove these artifacts, a discriminating factor has been designed to be the product of amplitude ratio, similarity, and power of beam. The effective discriminator significantly gets rid of most of the artifacts and leads to a very trustful imaging rupture model. All the subevents are imaged inside of the Longmen Shan fault zone. With the focal mechanism, the waveforms can be calibrated by the radiation pattern. There are two high-frequency releasing peaks during the earthquake. One is located at the epicenter nearby Yingxiu town, and the other one is nearby Beichuan county. Thus, the two high-frequency bursts are consistent with the most severe damaging zone at the surface. The good coincidence indicates the back-projection imaging results can provide strong constraints on the surface damaging estimation, greatly facilitating the earthquake response.

Due to the limited azimuthal coverage of the single generalized array, it is difficult to resolve the weak branch of the rupture of a bilateral earthquake using the relative back-projection method. To solve this issue, the fault plane of the 2010 bilateral Chilean earthquake is divided into two segments. These two segments are first back-projected separately and then combined together to analyze the rupture properties of the earthquake. Thus, more subevents on the southern weak branch are resolved. On the northern branch, the rupture is separated into two segments by an aseismic zone, where very little seismic energy is released.

The 2011 Mw 9.0 Tohoku earthquake is the earthquake recorded by unprecedentedly enormous number of instruments like strong motion stations, GPS campaigns, and tidal gauges. We carried out the relative back-projection method to image high-frequency sources of the earthquake using the direct *P* waves at the European seismic network. Using data with different frequency bands, the imaging results show that the high-frequency sources are located deeper and deeper as the frequency of the data used increases. We have demonstrated that this kind of frequency-dependent rupture feature cannot be resulted in by the relative back-projection method itself. Thus, the frequency-dependent feature is real and depends on the fault frictional property varying with depth. Meanwhile, the back-projection imaging results show that the earthquake is composed of three rupture segments with different rupture velocities. These rupture segments may be related to the uneven distributions of *P* wave, *S* wave, and Poisson's ratio on the causative fault. Thus, the heterogeneous structure or stress status of the fault probably controls the rupture properties of the earthquake. In addition, besides the typical rupture along the strike, there exist the evident ruptures along the dip. This is understandable due to the rupture of so large earthquake on the two-dimensional fault.

The limitation of the single array has been revealed in the research on the 2010 bilateral Chilean earthquake. In that study, we divided the fault plane into two segments back-projected separately since there is only one array, USarray transportable array, within a proper epidistance range. Alternatively, multiple arrays, located at different azimuths, can be applied to conduct the back-projection imaging to the 2012 Mw 8.6 Sumatra offshore earthquake. The multi-array back-projection imaging results show that the earthquake ruptures three separate, conjugate faults during 120 s. The rupture on the second causative fault first extends NNEward and then propagates SSWward. The rupture velocities on the two segments dramatically differ from each other; this could be caused by the difference of the rupture depth. There must be stress interaction among the causative faults. The Coulomb stress changes induced by the failure of the first fault triggers the second fault, and the dynamic stress changes induced by the passage of the SH wave or Love waves from the second fault triggers the third causative fault.

7.3 Perspectives

The relative back-projection method (RBPM) used in this book has been proofed both in theory and in applications to four large earthquakes. In general, this method can be effectively used to study the earthquake source. However, there exist some issues in applications of the RBPM. For instances, the reflected waves (*pP* and *sP*) have larger amplitudes than the direct P waves for the arrays AU and Fnet. For the 2010 Chilean earthquake, only the USarray, located north of the epicenter, is available to be utilized to imaging the rupture, which leads to ineffectiveness to resolve the rupture on the southern, weak branch.

To avoid the interference from depth phases, these phases can be removed but left the direct *P* waves. To realize this, the waveforms need to be cross-correlated with the theoretical Green's function. Next, the resultant cross-correlation function trains are used to conduct the back-projection imaging. However, the one-dimensional velocity model used to calculate the Green's function is oversimplified. In the future research, three-dimensional velocity model can be used to calculate the Green's function, which requires more computational resources. Alternatively, smaller aftershocks with a similar focal mechanism to the mainshock can be taken advantage as the empirical Green's function if there are enough aftershocks distributed on the fault plane.

For lack of the data, this can be solved when arrays composed of the oceanic observation seismometers that are available. In the future, the oceanic bottom seismometers will be deployed all over the sea floor; this will solve the issue that there lack data at some azimuths.

Nowadays, the waves used in the back-projection imaging are only one phase such as the direct *P* waves, *S* waves or surface waves. Besides one phase, maybe multiple phases can be combined together to provide much better constraints on the earthquake source just like the applications of waves in the finite fault modeling. In addition, the data widely used are teleseismic. Although the data are more complex with epidistance within a regional range, the regional or local data can be used in the back-projection imaging as well because the more detailed rupture information is included in these data.

There does exist frequency-dependent feature in the large earthquakes. All the data with different frequency bands such as the geodetic displacements, seismic data, and tsunamic water levels could be used to obtain the rupture of the earthquakes. Meanwhile, multiple approaches are used together to process these data. Thus, the earthquakes can be investigated using multiple data with different frequency bands.

The rupture processes of earthquakes are controlled by the frictional properties, stress status, fault strength, and fluid involvement. The investigation of the earthquake source is to reveal what roles these factors play in the occurrence of the earthquake. The more the earthquakes one studies, the deeper one understands the earthquake source. Eventually, one will have enough knowledge to reveal all the enigma of the earthquakes and will have the ability to predict the earthquakes someday.

References

Aki, K. 1984. Asperities, barriers, characteristic earthquakes and strong motion prediction. *Journal of Geophysical Research* 89 (B7): 5867–5872.

Allmann, B.P., and P.M. Shearer. 2007. A high-frequency secondary event during the 2004 Parkfield earthquake. *Science* 318 (5854): 1279–1283.

Capon, J. 1969. High-resolution frequency-wavenumber spectrum analysis. *Proceedings of the IEEE* 57 (8): 1408–1418.

Das, S., and K. Aki. 1977. Fault plane with barriers: A versatile earthquake model. *Journal of Geophysical Research* 82 (36): 5658–5670.

Dieterich, J.H. 1979. Modeling of rock friction: 1. Experimental results and constitutive equations. *Journal of Geophysical Research: Solid Earth (1978–2012)* 84 (B5): 2161–2168.

Kiser, E., and M. Ishii. 2011. The 2010 Mw 8.8 Chile earthquake: Triggering on multiple segments and frequency-dependent rupture behavior. *Geophysical Research Letters* 38 (7).

Lay, T., and H. Kanamori. 1981. An asperity model of large earthquake sequences. *Maurice Ewing Series* 4: 579–592.

Lay, T., H. Kanamori, and C.J. Ammon, et al. 2012. Depth-varying rupture properties of subduction zone megathrust faults. *Journal of Geophysical Research: Solid Earth (1978–2012)* 117: B04311.

Madariaga, R. 1977. High-frequency radiation from crack (stress drop) models of earth- quake faulting. *Geophysical Journal of the Royal Astronomical Society* 51 (3): 625–651.

Madariaga, R. 1983. High frequency radiation from dynamic earthquake fault models. *Annales Geophysicae* 1 (1): 17–23.

McLaskey, C.G., A. M. Thomas, S. D. Glaser, and R. M. Nadeau. 2012. Fault healing promotes high-frequency earthquakes in laboratory experiments and on natural faults. *Nature* 491 (7422): 101–104.

Roten, D., H. Miyake, and K. Koketsu. 2012. A Rayleigh wave back-projection method applied to the 2011 Tohoku earthquake. *Geophysical Research Letters* 39: L02302.

Rudnicki, J.W., and H. Kanamori, 1981. Effects of fault interaction on moment, stress drop, and strain energy release. *Journal of Geophysical Research: Solid Earth (1978–2012)* 86(B3): 1785–1793.

Schmidt, R. 1986. Multiple emitter location and signal parameter estimation. *Antennas and Propagation, IEEE Transactions on* 34 (3): 276–280.

Scholz, C.H. 1998. Earthquakes and friction laws. *Nature* 391 (6662): 37–42.

Yao, H., P. Gerstoft, P.M. Shearer, and C. Mecklenbrauker. 2011. Compressive sensing of the Tohoku-Oki Mw 9.0 earthquake: Frequency-dependent rupture modes. *Geophysical Research Letters* 38: L20310.

Yao, H., P.M. Shearer, and P. Gerstoft. 2013. Compressive sensing of frequency-dependent seismic radiation from subduction zone megathrust ruptures. *Proceedings of the National Academy of Sciences* 10 (12): 4512–4517.

Yue, H., T. Lay, and K.D. Koper. 2012. En echelon and orthogonal fault ruptures of the 11 April 2012 great intraplate earthquakes. *Nature* 490 (7419): 245–249.

FSC
www.fsc.org
MIX
Papier | Fördert
gute Waldnutzung
FSC® C083411

Kolibri 360 GmbH
Ferdinand-Jühlke-Straße 7
99095 Erfurt, Deutschland
produktsicherheit@kolibri360.de